## Praise for *Five Past Midnight in Bhopal*

'The true majesty of survival' Professor David Bellamy

'I hope your book significantly raises the safety consious-ness of managers in hazardous industries to a level where they can see the place that slippery slope is taking them' Warren Woomer, last American manager of the Bhopal plant

'A book that celebrates the spirit of human endeavour against all odds. A triumphant book' *India Today*, India

'A historical fresco full of love, heroism, faith and hope. A deeply moving book' *La Voz de Galicia*, Spain

'A sharp and stunning document you will not be able to put down' *Panorama*, Italy

'This book reveals in an exemplary way, the criminal under-takings of a major industrial group, in this case Union Carbide, which not only didn't measure the risks ran by surrounding populations, but also managed to escape all punishments' *Le Figaro*, France

# FIVE PAST MIDNIGHT IN BHOPAL

Dominique Lapierre
and Javier Moro

*Translated from the French by*
*Kathryn Spink*

Scribner

First published in Great Britain by Scribner, 2002
An imprint of Simon & Schuster UK Ltd
A Viacom Company

1   3   5   7   9   10   8   6   4   2

Simon & Schuster UK Ltd
Africa House
64–78 Kingsway
London WC2B 6AH

www.simonsays.co.uk

Simon & Schuster Australia
Sydney

A CIP catalogue record for this book is available from the British
Library

Hardback ISBN 0-7432-2034-X
Trade Paperback ISBN 0-7432 3088 4

Typeset by Palimpsest Book Production Limited,
Polmont, Stirlingshire
Printed and bound in Great Britain by
Omnia Books Limited, Glasgow

All photographs are from the authors' collection except: p.4–5, coll.
Eduardo Muñoz; p.6 (top), coll. Zahir ul-Islam; p.6 (bottom), coll. Niloufar
Khan; p.8 (top), coll. John Luke Couvaras; p.13–14 (up), p.15–16 (top,
left bottom), coll. Jamaini.

To the heroes of the Orya Bustee, of Chola
and of Jai Prakash Nagar

# Acknowledgements

First and foremost we would like to express our immense gratitude to our wives, Dominique and Sita, who shared every moment of our long and difficult research and who were our irreplaceable helpers in the preparation of this work.

Heartfelt appreciation to Colette Modiano, to Paul and Manuela Andreota, to Pascaline Bressan and Michel Gourtay, to Mari Carmen Doñate, Eugenio Suarez and Antonio Ubach who spent long hours correcting our manuscript and gave us their encouragement.

A very special thank you to Antoine Caro for his exceptional assistance with the preparation of this book, as well as to Pierre Amado for his valuable advice on India.

This book is the fruit of patient research both in the United States and in India. In the United States we would like particularly to thank engineer Warren Woomer and his wife Betty who made us welcome in their charming house in South Charleston, enabling us to reconstruct the happy years when Warren was in charge of the Bhopal factory. Similarly we would like to thank engineer Eduardo Muñoz and his wife Victoria for our innumerable meetings in San Francisco and at their villa in Sausalito, in the course of which we were able to reconstruct, almost day by day, the adventure of the establishment of a high tech pesticide plant in the heartland of India, and the fight Muñoz had to limit its size and the dangers involved.

Again in the United States, we would like to thank Halcott P. Foss and engineers Jean-Luc Lemaire and William K. Frampton, for having opened wide the doors to the Institute factory, the Bhopal plant's elder sister, where Sevin is still produced from deadly methyl isocyanate. Additional thanks go to Jean-Luc Lemaire and René Crochard for the illuminating explanations that facilitated the writing of the technical parts of our book. We include in this American tribute Ward Morehouse and David Dembo, who from their small East River office in New York conduct an unrelenting struggle to let the truth about the Bhopal disaster be known and who generously gave us access to their precious archives. Equally we would like to express our gratitude to Kathy Kramer for having placed at our disposal documentation concerning the Boyce Thompson Institute in Yonkers where the Sevin that was to wipe out insects ravaging the harvests of peasants throughout the world was invented.

Among all the Indian engineers who took part in the adventure of Bhopal's 'beautiful plant', our gratitude is due primarily to Kamal Pareek for the entire days we spent together, reconstructing in every little detail the extraordinary hope that the setting up of the Bhopal factory brought with it, its subsequent slow agony and the eventual catastrophe. Grateful thanks also go to engineers Umesh Nanda and John Luke Couvaras who patiently shared their memories with us and entrusted numerous unedited documents to us. We would similarly like to express our gratitude to Jagannathan Mukund who was the factory's last managing director and who allowed us to bombard him with questions for three days in his Conoor property in the Nilgiri hills in southern India.

Naturally a very large part of our research was conducted in Bhopal itself, where the assistance of Satinath Sarangi and his team of record keepers from the Sambhavna Trust was indispensable to us, as were the generous help and hospitality of Farah Khan and her mother Nilufar Khan; Begum Rashid, Bano and Yadar Raachid, Sonia and Nader Raachid, Uzzafar Khan as well as Mr and Mrs Balthazar de Bourbon, Enamia, Kamlesh Jamaini, the chronicler Nasser Kamal, Manish Mishra and Dr Zahir ul-Islam who helped us uncover the secrets of the culture and legendary past of their beautiful city.

We wish to thank also Mr Digvijay Singh, the Chief Minister of Madhya Pradesh, for his warm reception, and all those who so generously helped us in the various aspects of our research. In alphabetical order: MM Shyam Babu, K. D. Ballal in Bangalore, Dr Bambhal, Sudeep Banerjee, Sajda Bano, Ahmed Bassi, Dr Bhandari, Praful Bidwai, N. M. Buch, Father Dennis Carneiro, Amar Chand, Dr Heeresh Chandra, T. R. Chouhan, S. P. Chowdhary, Mr Chughtai, Deena Dayalan and the staff of *The Other Media*, Mr Diwedi, Dr Banu Dubey, R. K. Dutta, Dr Deepak Gandhe, Brigadier Garg, Subhash Godane, V. P. Gokhale from Eveready, Ahsan Hussain, Santosh Katiyar, Rehman Khan, Colonel Gurcharan Singh Khanuja, Rajkumar Keswani, Dr Loya, Dr N. P. Mishra, Dr Nagu, Shekil Qureshi, Ganga and Dalima Ram, Dr Rajanarayan, Salar, Dr Sarkar, Dr Satpathy, Arvind Shrivastava, V. N. Singh, Commissioner Ranjit Singh, S. K. Trehan, Dr Trivedi, Dr Varadajan, Mohan Lal Varma, Reverend Timothy Wankhede, Sister Christopher Wheelan.

Union Carbide's management failed to respond to any of our requests for interviews and information.

By contrast, we are grateful to the Rhône Poulenc company, which took over the proprietorship of the Institute factory in the United States, and to its director for Agro international public relations, Georges Santini, for having generously received us both in Institute and at the research department in Lyon. We include in our appreciation Christine Giulani, in charge of public relations for Dow Agro Sciences, for the warm welcome provided at the Letcombe Regis laboratories in Great Britain.

We want to thank also our friends who made our travels and stays in India so productive and pleasant: MM Sanjay Basu and all the staff at *Far Horizon*, Ranvir Bhandari, Audrey Daver, Bharat Dhruv, Madan Kak and the whole staff of TCI, Sanjiv Malhotra, Sunil Mukherjee, Gilbert Soulaine and Gilles Renard.

We address our special gratitude to those who help us so generously in our humanitarian work: their excellencies the ambassadors Bernard de Montferrand and Kanwal Sibal, Mary Allizon, Rina and Takis Anoussis, David Backler and the Foundation Marcelle & Jean Coutu, Otto Barghezi, Jamshed Bhabha; Drs Françoise Baylet-Vincent, Angela Bertoli, Henri-Jean Philippe and their benevolent friends of the organizations 'Gynécologie sans frontières' and 'Pathologie, Cytologie et Développement'; Lon and Dick Behr, Nicolas Borsinger and the Foundation ProVictimis; Pierre Ceyrac, Kathryn and John Coo, Gaston Dayanand, Peter and Richard Dreyfus, Behram and Mani Dumasia, Catherine and David Graham, Priti Jain, Mohammed Kamruddin and the whole team of UBA, Adi and Jeroo Katgara, Ashwini and Renu Kumar, François Laborde and the whole team of HSP, Ila Lumba, Michèle Migone and all the Friends of Italy, Christina Mondadori and the Foundation Benedetta d'Intino, Aman Nath, Aloka Pal; Sabitri Pal, Shirin Paul, Mohammed Abdul Wohab and the whole staff of SHIS, Gaston Roberge, June and Paul Shorr; James Stevens and the whole team of Udayan; Sukhesi Didi and the whole staff of Belari, Ratan Tata, chairman of the Tata Group, Suzanne and Alexander Van Meerwijk, Francis Wacziarg, Harriet and Larry Weiss; and all those who prefer to remain anonymous.

We could not have written this book without the enthusiastic faith of our publishers. Our warm thanks to Leonello Brandolini, Nicole Lattès and Antoine Caro in Paris; Carlos Reves and Berta Noy in Barcelona; Gianni Ferrari and Joy Terekiev in Milan; Larry Kirshbaum and Jessica Papin in New York; Shekhar and Poonam Malhotra in Delhi; Helen

Gummer and Katharine Young in London; and finally to our friend and translator Kathryn Spink, herself the author of remarkable works on Mother Teresa, Brother Roger of Taizé, Little Sister Magdeleine of Jesus and Jean Vanier.

# Contents

## PART ONE
## A NEW STAR IN THE INDIAN SKY

PART TWO

A NIGHT BLESSED BY THE STARS

PART THREE

THREE SARCOPHAGI UNDER THE MOON

# Letter to the Reader

One day I met a tall Indian in his forties, with a red bandanna round his head and hair knotted in a braid at the back of his neck. The brightness of his smile and the warmth of his expression made me realize immediately that this was a man with compassion for the poor. Having heard that the second 'City of Joy' dispensary-boat had just been launched in the Ganges Delta to bring medical aid to the inhabitants of fifty-four islands, he wanted to ask for my help.

Right after he got the news of a deadly chemical accident in the city of Bhopal, Satinath Sarangi, 'Sathyu' as he is called, rushed to the rescue of the survivors of the worst industrial disaster in history, a massive leak of toxic gases, which, on the night of the second to the third of December 1984, killed between 16,000 and 30,000, and injured around 500,000. Sathyu decided to dedicate his whole life to the victims. Since 1995, he has been running a non-governmental, non-political and non-religious organization, tirelessly caring for the poorest and most neglected men, women and children affected by the gas.

Sathyu wanted to ask me to finance the creation and equipment of a gynaecological clinic to treat underprivileged women who, sixteen years after the tragedy, were still suffering from its dreadful effects.

I had a vague recollection of the tragedy but, in all my fifty years of roving about this vast country, I had never visited the magnificent capital of Madhya Pradesh.

I went to Bhopal. What I found there gave me what was probably one of the strongest shocks of my life. With the help of my royalties and the generosity of readers of *The City of Joy, Beyond Love* and *A Thousand Suns*, we were able to open the gynaecological clinic. Today it takes in, treats and cures hundreds of women whom the town's hospitals had abandoned to their fate.

Above all, however, the experience pointed me in the direction of one of the most enthralling subjects of my career as a journalist and writer: why and how could such a monumental accident take place? Who were the people who initiated it, those involved in it, the victims of it, and finally, who benefited from it?

I asked the Spanish writer Javier Moro, author of *The Mountains of the Buddha*, a moving book on the tragedy of Tibet, to join me in Bhopal. Our research went on for three years. This book is the fruit of it.

Dominique Lapierre

# The City of Bhopal

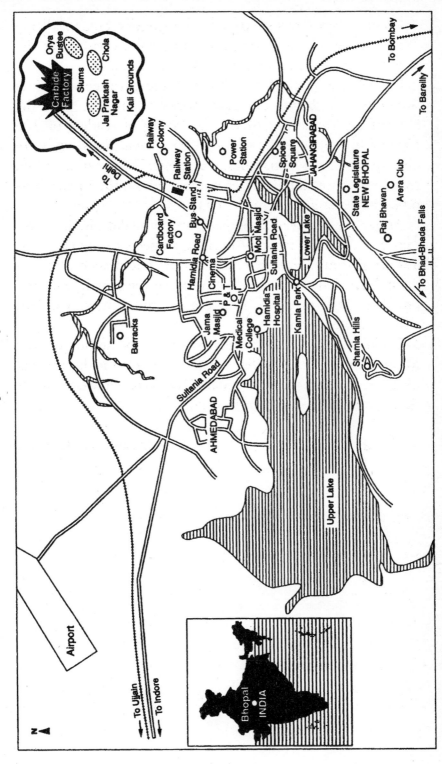

*Concern for man himself and his safety must always form the chief interest of all technical endeavours.*

*Never forget this in the midst of your diagrams and equations.*

Albert Einstein

## PART ONE

# A NEW STAR
# IN THE INDIAN SKY

# 1

## *Firecrakers that kill, cows that die and insects that murder*

Mudilapa. One of India's 1,500,000 villages and probably one of the poorest in a country the size of a continent. Situated at the foot of the remote hill region of Orissa,
it comprised some sixty families belonging to the Adivasi
community, descendants of the aboriginal tribes which populated India over 3000 years ago and which the Aryans from
the north drove back into the less fertile mountainous areas.

Although officially 'protected' by the authorities, the
Adivasis remained largely beyond the reach of the development programmes which were trying to improve the plight
of the Indian peasants. Deprived of land, the inhabitants of
the region had to hire out their hands to make a living for
their families. Cutting sugar cane, going down into the
bauxite mines, breaking rocks along the roads, no task was
too menial for those disenfranchised by the world's largest
democracy.

'Goodbye wife; goodbye children; goodbye father, mother,
parrot. May God watch over you while I'm away!'

At the beginning of every summer, when the village lay
cloaked in a leaden and blazing heat, a lean, dark-skinned
muscular little man would bid farewell to his family before

setting off with his bundle on his head. Thirty-two-year-old Ratna Nadar was embarking on a strenuous journey: three days of walking to a palm grove on the shores of the Bay of Bengal. Because of the strength in his arms and hamstrings he had been taken on by a *tharagar*, an agent who travels about recruiting labourers. Work in palm groves required an unusual degree of agility and athletic strength. Men had to climb, bare-handed and without a safety harness, to the top of date palms as tall as five-storey houses, to make a nick in the axil of the palm and collect the milk secreted from the heart of the tree. These acrobatic ascents earned Ratna Nadar and his companions the nickname 'monkey-men'. Every evening the manager of the enterprise would come and take their precious harvest and transport it to a confectioner in Bhubaneshwar.

Ratna Nadar had never actually tasted this delicious nectar. But the 400 rupees he earned from a season spent risking life and limb enabled him to feed the seven members of his family for several weeks. As soon as his wife had wind of his return, she would light an incense stick before the image of Jagannath, which decorated one corner of the hut, and thus gave thanks to the 'Lord of the Universe', a manifestation of the Hindu god Vishnu adopted by the Adivasis. Sheela was a frail but spirited woman with a ready smile. The plait down her back, her almond-shaped eyes and rosy cheeks made her look like a Chinese doll. There was nothing very surprising about that: her ancestors belonged to an aboriginal tribe, originally from Assam, in the far north of the country.

The Nadars had three children. The eldest, 8-year-old Padmini, was a delicate little girl with long dark hair tied in two plaits. She had inherited Sheela's beautiful, slanting eyes and her father's determined profile. The small gold ring, which she wore, as tradition dictated, through the ala of her

4

nose, enhanced the brightness of her face. Getting up at dawn and going to bed late, Padmini helped her mother with all the household chores. She had helped to raise her two brothers, 7-year-old Ashu and 6-year-old Gopal, two tousle-haired little rascals more inclined to chase lizards with a palm frond than go and fetch water from the village water-hole. Ratna's parents also shared the Nadars' home: his father Prodip, whose gaunt face was traversed by a thin, grey moustache, and his mother Shunda, already wrinkled and bent.

Like tens of millions of other Indian children, Padmini and her brothers had never been anywhere near a school blackboard. All they had been taught was how to survive in the harsh world into which the gods had ordained they should be born. And, like all the other occupants of Mudilapa, Ratna Nadar and his family lived on the look-out for any opportunity to earn the odd rupee. One such opportunity arose each year at the beginning of the dry season when the time came to pick the various leaves used to make *bidis*, the slender Indian cigarettes with the tapered tips.

For six weeks, along with most of the other villagers, Sheela, her children and their grandparents, would set off each morning at dawn for the forest of Kantaroli. There the people would invade the undergrowth like a swarm of insects. With all the precision of robots, they would detach a leaf, place it in a canvas haversack and repeat the same process over and over again. Every hour, the pickers would stop to make up bunches of fifty leaves. If they hurried, they could generally manage to produce eighty bunches a day. Each bunch was worth 30 paisa, not quite one penny, or the price of two aubergines.

During the first days, when the picking went on at the edge of the forest, young Padmini would often manage to make as many as 100 bunches. Her brothers Ashu and Gopal were not quite as dexterous at pinching the leaves off. But between the six of them, the children, their mother and their grandparents brought back nearly 100 rupees each evening, a small fortune for a family used to surviving for a whole month on far less.

⁂

One day, word went round Mudilapa and the surrounding villages that a cigarette and match factory had recently been set up in the area and that children were being taken on as labour. Of the hundred billion matches produced annually in India, many were still made by hand, primarily by children, whose little fingers could work wonders. The same applied to rolling *bidis*.

The opening of this factory created quite a stir amongst the inhabitants of Mudilapa. There were no lengths to which people would not go to seduce the *tharagar* whose job it was to recruit the workforce. Mothers rushed to the *mohajan*, the village usurer, and pawned their last remaining jewels. Some sold their only goat. And yet the jobs they sought for their children were harsh in the extreme.

'My truck will come by at four every morning,' the *tharagar* announced to the parents of the children he had chosen. 'Anyone who is not outside waiting for it had better look out.'

'And when will our children be back?' Padmini's father gave voice to all the other parents' concern.

'Not before nightfall,' the *tharagar* responded curtly.

Sheela saw an expression of fear pass over Padmini's face. She sought at once to bolster her resolve.

'Padmini, think what happened to your friend Banita.'

Sheela was referring to the neighbours' little girl whose

6

parents had just sold her to a blind man so they could feed their other children. There was nothing particularly unusual about the arrangement. Sometimes in all innocence, parents entrusted their daughters to pimps, thinking they were going to be employed as servants or in a workshop.

It was still pitch dark when the truck horn sounded the next morning. Padmini, Ashu and Gopal were already outside, waiting, huddled together. They were cold. Their mother had got up even earlier to prepare a meal for them: a handful of rice seasoned with a little dhal[1], two chapattis[2] each and a chilli pepper to share, all wrapped in a banana leaf.

The truck stopped outside a long, open, tiled shed, with a baked earth wall at the back and pillars to support the roof at the front. It was not yet daybreak and kerosene lamps scarcely lit the vast building. The foreman was a thin, overbearing bully of a man, wearing a collarless shirt and a white loincloth.

'In the darkness, his eyes seemed to blaze like the embers in our *chula*[3],' Padmini recalled.

'All of you sit down along the wall,' he ordered.

Then he counted the children and split them into two groups, one for cigarettes, the other for matches. Padmini was separated from her brothers and sent to join the *bidi* group.

'Get to work!' the man in the white loincloth commanded, clapping his hands.

His assistants then brought trays laden with leaves like those Padmini had picked in the forest. The oldest assistant squatted down in front of the children to show them how

---

[1] Lentil purée which is the main source of vegetable protein in India.
[2] Corn pancake.
[3] A small, rudimentary oven.

to roll each leaf into a little funnel, fill it with a pinch of shredded tobacco and bind it with a red thread. Padmini had no difficulty in imitating him. In no time at all she had made up a packet of *bidis*. 'The only thing I didn't like about it was the pungent smell of the leaves,' she confided. 'To get through the pile of leaves in front of us, we found it best to concentrate on the money we'd be taking home.'

Other workmen deposited piles of tiny sticks in front of the children assigned to making matches.

'Place them one by one in the slots of this metal support,' the foreman explained. 'Once it's full, turn it round and dip the ends of the sticks in this tank.'

The receptacle contained molten sulphur. As soon as the tips had been dipped and lifted out again, the sulphur solidified instantly.

Padmini's younger brother surveyed the steaming liquid apprehensively.

'We'll burn our fingers!' he said anxiously, and loudly enough for the foreman to hear.

'You little idiot!' the man retorted. 'I told you, you only immerse the end of the wooden sticks, not the whole thing. Have you never seen a match?'

Gopal shook his head. But his fear of being burned was nothing compared with the real risk of being poisoned by the toxic fumes coming off the tank. It was not long before some of the children began to feel their lungs and eyes burning. Many of them passed out. The foreman and his assistants had to revive them by slapping them round the face and dousing them with buckets of water. Those who fainted again were mercilessly expelled from the factory.

'Shortly after our arrival, a second shed was built to house a work unit to make firecrackers,' Padmini recounted. 'My brother Ashu was assigned to it together with about twenty other boys. After that I only saw him once a day, when I took him his share of the food our mother had prepared for us. The foreman would ring a bell to announce the meal break. Woe betide any of us who were not back in our places by the second bell. The boss would beat us with the stick he carried to frighten us and make us work faster and faster. Apart from that short break, we worked without interruption from the time we arrived until nightfall when the truck would take us home again. My brothers and I were so tired we would throw ourselves onto the *charpoy*[4] without anything to eat and fall asleep straight away.'

A few weeks after the opening of the firecracker unit, tragedy struck. Suddenly Padmini saw a huge flame blazing in the shed where her brother Ashu was working. An explosion ripped away the roof and wall. Boys emerged screaming from the cloud of smoke. They were covered in blood. Their skin was hanging off them in shreds. The foreman and his assistants were trying to put out the fire with buckets of water. Padmini rushed frantically in the direction of the blaze, shouting her brother's name, but there was no answer. She was running about in all directions when she stumbled. As she fell, she saw a body on the ground. It was her brother. His arms had been blown off in the blast. 'His eyes were open as if he were looking at me, but he wasn't moving,' she said. Ashu was dead. Around him lay other little injured bodies. 'I picked myself up and went and took my other brother's hand. He had taken refuge in a corner of the match shed. I sat down beside him, held him tightly in my arms, and together we wept in silence.'

※

---

[4] Literally 'four legs'. A bed made out of rope strung across a wooden frame.

One month after this accident, a uniformed official from the Orissa Department of Animal Husbandry made his appearance in Mudilapa. Driving a jeep equipped with a revolving light and a siren, he was the first government representative ever to visit the village. He summoned the villagers with a loudhailer and the entire population assembled round his jeep.

'I have come to bring you great news,' he declared, caressing the microphone with fingers covered in rings. 'In accordance with her policy of helping our country's most underprivileged peasants, Indira Gandhi, our Prime Minister, has decided to give you a present.' With some amusement the man took in the astonishment clearly visible on the faces of those present. Waving a hand at random in the direction of one of them, he enquired:

'You, do you have any idea what our mother might want to give you?'

Ratna Nadar, Padmini's father, hesitated.

'Perhaps she wants to give us a well,' he ventured.

Already, the man in uniform had turned to someone else.

'And you?'

'She's going to make us a proper road.'

'And you?'

'She wants to provide us with electricity.'

'And you? . . .'

In less than a minute, the government envoy was in a position to assess the state of poverty and neglect in the village. But the purpose of his visit was nothing to do with any of these pressing needs. Heightening the suspense with a protracted silence, at last he continued:

'My friends, I've come to inform you that our beloved Indira has decided to give every family in Mudilapa a cow.'

'A cow?' repeated several stupefied voices.

'What are we going to feed it on?' someone asked anxiously.

10

'Don't you worry about that,' the visitor went on, 'Indira Gandhi has thought of everything. Every family is to receive a plot of land on which you'll grow the fodder you need for your animal. And the government will pay you for your labours.'

It was too good to be true.

'The gods have visited our village,' marvelled Padmini's mother, who was always ready to thank Heaven for the slightest blessing. 'We must offer a *puja*[5] at once.'

The government envoy carried on with his speech. He spoke with all the grandiloquence of a politician come to dispense gifts before an election.

'Don't go, my friends, I haven't finished! I have an even more important piece of news for you. The government has made arrangements for each one of your cows to give you a calf from semen taken from specially selected bulls imported from Great Britain. Their sperm will be brought to you regularly from Bombay and Poona by government vets who will themselves carry out the insemination. This programme should produce a new breed in your region, capable of yielding eight times more milk than local cattle. But take note that to achieve this result, you will have to undertake never to mate your cow with a local bull.'

The bewilderment on the faces of the onlookers was replaced with joy.

'Never before have we had a visit from a benefactor like you,' declared Ratna Nadar, sure that he was relaying the gratitude of them all.

The day the herd arrived, the women dug out their wedding saris and festival veils from the family coffers just

---

[5] A ceremonial offering in front of the altar of a god.

as they did for the Diwali[6] or Dassahra[7] celebrations. All night long they danced and sang round the animals, who joined in with a concert of mooing. The Nadars named their cow after Lakshmi, the goddess of wealth, to whom the Adivasis were as fervently devoted as Hindus.

Just as the government envoy had announced, a few weeks later vets arrived in Mudilapa. They came bearing fat syringes to inseminate the cows with British sperm. Ten moons later, in the yard outside every hut in the village, a calf made its entry into the world. But the villagers' joy lasted only one night. Not one of the young calves managed to get to its feet and suckle from its mother. Sheela tried in vain to induce the starving newborn to drink a little milk out of a coconut shell. One after another all the calves died. It was a disaster.

'I'm going to take Lakshmi to a local bull,' Ratna Nadar informed his family one morning.

'It's because their fathers aren't from round here that our calves died,' he said.

His neighbour decided to do the same, but the attempt proved fruitless. The government agents had taken their own precautions. To prevent the peasants from having their cows inseminated by a local bull, they had had them all castrated.

❋

The inhabitants of Mudilapa took heart once more when they saw the young shoots they had sown for their cows on the half-acre allotted to them by the government sprouting

---

6 Diwali, the festival of lights celebrated with an explosion of fireworks and firecrackers, is the most joyous festival in the Hindu calendar.
7 Dassahra, the tenth day of the festival of Durga celebrating the goddess's victory over the buffalo demon of ignorance.

from the ground. At least they would be able to feed their cattle. Every morning Ratna Nadar took his family to the field to watch over the welfare of the future harvest. One day, they noticed that the grass had changed colour. It had turned grey. It couldn't be for want of water because the soil was still damp from the last rains. On careful examination of the stems, Ratna discovered that they were infested with black aphids that were devouring the outer layer and sucking up the sap. The other peasants found the same. Calamity had struck Mudilapa. Was Jagannath angry? The Nadars and their neighbours went to ask the village priest to offer a *puja* to the great god in order that their fields might be restored to health. Without fodder, their cows would die. The old man with his shaven head traced a circle around a few shoots and began to dance, chanting ritual prayers. Then he sprinkled them with *ghee*, clarified butter, and set fire to them one by one.

But Jagannath refused to hear. Consumed by aphids, the Nadars' fodder died in a matter of days. It was September and they would not be able to sow again until the following spring. Soon their cow was reduced to skin and bone. The region's cattle merchants got wind of the catastrophe. Like vultures they came to buy the animals up dirt-cheap while they were still alive. The Nadars had to resign themselves to letting Lakshmi go for fifty rupees.

The sale enabled them to hold out for a few more weeks. When the elderly Shunda, the grandmother who kept the family savings wrapped up in a handkerchief, had got out her last few coins, Ratna gathered his family round him.

'I'm going to the *mohajan*,' he declared. 'I shall give him the fields as security for his lending us something to live on until next seed-time. This time we'll sow corn and lentils. And we'll find a way of preventing those cursed little creatures from devouring our harvest.'

'Ratna, father of my children,' Sheela interrupted timidly, 'I've hidden it from you until now so as not to worry you, but you must know that we no longer have a field. One day when you were away, working in the palm grove, the government people came and took back all the plots of land they found with no crops on them. I tried to tell them that insects had eaten what we had planted, but they would hear none of it. The man in charge shouted "You're useless!" and tore up the papers they gave us when they brought us the cows.'

This revelation plunged the family into consternation. No one could muster the strength to say anything. This time the Nadars seemed to have come to the end of their tether. Then a child's voice rang out in the overheated hut.

'I'll go back to rolling *bidis*,' declared Padmini.

✳

Her courageous offer was not to be taken up. A few days later, an unknown *tharagar* arrived in Mudilapa. He had been sent by the Madhya Pradesh Railroad to recruit a workforce to double the railway lines into the station in Bhopal, the state capital.

'You could earn as much as thirty rupees a day,' he told Ratna Nadar, carefully examining the date palm climber's muscles with a professional eye.

'What about my family?' asked Nadar.

The *tharagar* shrugged his shoulders.

'Take them with you! There's plenty of room in Bhopal!' He counted the number of people in the hut. 'There you go. Six train tickets for Bhopal,' he said, taking six small squares of pink paper out of his *lunghi*[8]. 'It's a two to three day

---

[8] Long cotton loincloth knotted at the waist.

journey. And on top of that, here's a fifty rupee advance on your first wages.'

The deal had been done in less than five minutes. There was nothing generous about this gesture on the part of the *tharagar*. The Adivasis were well known to be as undemanding as they were exploitable.

⁂

The Nadar family's exodus posed hardly any problems. Apart from a few tools, linens and household utensils and Mangal, the irrepressible parrot with his red and yellow plumage, they had no possessions. The next monsoon storms would demolish the hut, unless some passing family took possession of it in the meantime.

One morning, just as Surya, the sun god, was casting his first pink rays over the horizon, the Nadars set off, with Ratna and his father, Prodip, leading the way. They all carried bundles on their heads. The small caravan, to which other Mudilapa families had attached themselves, left behind it a cloud of dust. Young Gopal, parrot cage in hand, pranced for joy at the prospect of adventure. Padmini, however, could not hold back tears. Before the road veered away to the north she looked back over her shoulder for one last time and bade farewell to the hut that had been her childhood home.

# 2

## The planetary holocaust wrought by armies of ravaging insects

The misfortune of the peasants from Mudilapa was just one tiny episode in a tragedy affecting the entire planet. The black aphids that had driven the Nadars from their land were among 850,000 varieties of insect which, since the dawn of humanity, have been stripping mankind of half our food supply. Many of the names give scant indication of the nature and magnitude of the disasters they cause. How, after all, would anyone ever suspect the diminutive creatures so lyrically christened 'oriental fruit moths', 'red-banded leaf rollers', 'rosy apple aphids', 'striped stem borers' or indeed 'white-backed planthoppers', of such capacity for destruction? The range of their weapons, the extreme diversity of their appearance, the flamboyance of their carapaces makes these parasites the most fabulous creatures in the Bestiary of God's imagination. The dazzling iridescence of some fruit-eating moths is reminiscent of glitteringly bejewelled apparel and quite unlike the hairy coat of the repulsive caterpillars that destroy cotton fields. Every species has its own method of survival. Some suck, like Mudilapa's Indian aphids. Others crush, devour, lick, ravage, pillage or bite. Then there are pulp-eaters, plant-eaters and wood-eaters. Some grind up vegetation with their mandibles, some suck it dry with a long proboscis, others lick it before sucking it up through a sheath encircling their tongue, yet others stab it with a dagger, then pump out the sap. Some nibble at leaves, gnaw them into crenellated shapes, scour them down or puncture them with little holes. Others invade the leaf canals and spread themselves through the veins. Dense foliage suddenly finds itself thus riddled with whitish spots, harbouring armadas of assailants the size of pinheads. All at once, healthy, vigorous plants find themselves covered with brownish powdery pustules, which cause them to wither and die.

The muscular, ballerina-like thighs of Mexican bean beetles enable them to jump from stem to stem like circus acrobats, while yellow stem borers haul themselves over leaves as laboriously as tortoises. The beetles that kill cereals are thread-like; the tipula that destroy vegetables look like mosquitoes engorged with blood. Moths with shimmering, scaly, double wings that live on lentils, hairy thrips that kill olive trees, scarlet acarius worms that are the terror of the orchard – all form part of a sinister, infinitely small jungle teeming with life.

Because of their unlimited capacity to adapt, these insects are found in any environment or latitude – as readily in the blazing sands of the African desert as under Arctic ice floes. Some have been responsible for many of humanity's worst catastrophes: the grasshoppers that plagued ancient Egypt, for example; the phylloxera aphid that wiped out the French vineyards at the end of the last century; the Colorado beetle that caused the Irish potato famine.

All these little creatures are unrelenting and unsparing of any crop. The same maggot can travel from ear to ear in fields of corn, barley, oats or rye. Stems punctured, roots consumed by the bulimia of millions of larvae, cereal crops are destroyed on a vast scale without producing a single grain. Rice, which makes up 60 per cent of the grain used as food for the world's population, is a prime target for these plunderers. Rice whorl maggots with their cylindrical bodies, pink semiloopers, biting spring-beetles, greyish-coloured, legless tipula, the army worm caterpillar, mayfly larvae, leafhoppers, millipedes, thread-worms – over a hundred species attack this single cereal. Among the most devastating is a moth called pyralis, meaning literally 'insect living in fire', whose long, greyish caterpillars dig tunnels in the ears of rice until they topple over. No less destructive are certain vicious little creatures armed with a sucking stylet and protected from predators by a thick carapace. As for weevils, they have probably gobbled up more rice, corn and potatoes than humankind has managed to consume since agriculture first began.

For thousands of years, man has been conducting a desperate war against the authors of this destruction. Texts from Ancient China, the Rome of antiquity and medieval Europe, all abound in extraordinary accounts of such battles. In the absence of any effective scientific means of warding off attack, our ancestors relied on magical and religious practices. Nepalese peasants put up notices in their paddy fields prohibiting insects from entry 'on pain of legal proceedings'. Less naïve but just as unrealistic, Roman peasants had pregnant women walk in circles round their fruit trees. Medieval Christians organized processions and novenas to counteract the cochylis and the corn and vine pyralis. Farmers in Venezuela beat the ears of their grain with belts in the hope that shock treatment would strengthen their plants' resistance to parasites. While Siamese farmers dotted their fields with eggshells pinned to sticks, those in Malaysia attached dead toads to bamboo poles to drive the white fly from their rice fields. Believing insect attacks to be the necessary consequence of sin in a divine and perfect creation, people in the year AD 1000 had no qualms about taking them to court. In Lausanne, in 1120, caterpillars were excommunicated. Five centuries later a court in Auvergne condemned other caterpillars to go and 'finish their wretched lives' in a place expressly provided for the purpose. Insects went on being brought to trial until 1830.

Fortunately, other more realistic countermeasures had been tried. By flooding their fields at certain times of the year, peasants in the south of India had managed to drown destructive insects in their millions. In Kenya and Mexico the simple idea of planting squares of maize as lures in the middle of other crops had saved vegetable and sorghum plantations. Elsewhere the use of predatory insects had won some splendid victories. Texts dating from the third century report that Chinese growers infested their lemon trees with ants, which ate vanessas, the richly coloured butterflies that devoured their orchards. Fifteen centuries later, the man dibles of a killer scarabaeid beetle saved the citrus plantations of California from the ravages of the Australian fly.

At the end of the nineteenth century vegetable-based materials such as nicotine or the pyrethrum flower, then mineral substances such as arsenic and copper sulphate, supplied peasants with new weapons, which they would soon call by the magic name of insecticides and later pesticides. With the discovery, in 1868, that spraying the arsenic-based dye 'Paris green' on cotton parasites had a guaranteed effect, the United States launched a frenzied commercialization of natural poisons. By 1910, the new American pesticide industry was worth more than 20 million dollars. To 'Paris green' were added lead arsenate based products. The First World War brought about explosive expansion in other directions. With German submarines preventing the importing of 'Paris green' and the war effort commandeering arsenic for the manufacture of munitions, flares and combat gas, insecticide producers turned to the chemical industry.

Only too delighted to find an outlet for petroleum by-products, chemists were quick to take up the challenge. Large European and American companies invested huge sums of money in research into synthetic molecules that could destroy predatory insects. Thus the period between the two world wars witnessed the advent of a string of chemical families, each with new capabilities for exterminating parasites. The goal appeared to be reached on the eve of the Second World War when Herman Mueller, a Swiss chemist trying to come up with an effective contact insecticide, discovered a molecule that seemed to meet his requirements. It bore the unwieldly name of Dichloro-Diphenyl Trichloroethane. Fortunately the Swiss scientist found a shorter, more convenient label. The insect world was to tremble: DDT had been born. This spectacular discovery earned its author the Nobel Prize for Physiology and Medicine, for DDT would bring about the mass extermination of malaria-carrying mosquitoes in the field of military operations, thereby saving the lives of hundreds of thousands of soldiers. At the end of the war this organic insecticide was put to the civilian use for which it had been invented. Field studies showed that it swiftly destroyed an extensive range of plant-eating insects, and thus immediately increased agricultural yield. Experiments carried out in New York and Wisconsin revealed that the yield from potato fields treated with DDT shot up by 60 per cent. The euphoria that greeted these gratifying results subsided when it was found that DDT also contaminated the earth, mammals, birds, fish, and even people. It was soon declared illegal in most Western countries. In Europe and the United States, legislation was introduced obliging pesticide manufacturers to respect the increasingly draconian protection and safety standards.

Under pressure from an impatient agricultural industry, they geared their research to finding products that would reconcile the destruction of insects with a level of toxicity tolerable to humanity and its environment. An extraordinary adventure was about to begin.

# 3

## A neighbourhood called Orya Bustee

After fifty-nine hours in the colourful congestion of an Indian train, the exiles from Mudilapa at last reached Bhopal, their journey's end. In the months that followed India's independence, this prestigious city had been made the capital of Madhya Pradesh, a state almost the size of France situated at the geographical heart of the country. Padmini Nadar and her family had remarked continuously on the beauty of the countryside they traversed, especially as they drew nearer to the city. Wasn't it in the deep, mysterious forests the train was passing through, that the god Rama and the Pandava brothers of Hindu mythology had taken refuge, and that Rudyard Kipling had set *The Jungle Book?* And wasn't it true that there were still tigers and elephants in that jungle? A few miles before their destination the railway had run past the famous caves of Bhimbekta, the walls of which were decorated with prehistoric aboriginal rock paintings.

The station where the immigrants from Orissa got off was one of those caravanserais swarming with noise, activity and smells, typical of India's large railway terminals. It dated back to the previous century. Not even the most colourful festival in Adivasi folklore could have given Padmini or her family an inkling of the celebrations staged in that station on 18 November 1884, its inauguration day. A British colonial administrator had proposed the idea of linking the ancient princely city to the rail network after a terrible drought had caused tens of thousands of local people to die of starvation, deprived of aid for want of communication lines. History has largely overlooked the name of the flamboyant Henry Daly who was responsible for giving Bhopal the most valuable asset an Indian town could then receive from its colonizers. A whole retinue of Britannic excellencies in braided uniforms studded with medals and all the local dignitaries in ceremonial costumes had come running at the invitation of the Begum, a slight woman hidden beneath the folds of a *burkah*[1], who ruled over the Sultanate of Bhopal. The festivities went on for three days and three nights. The local people had gathered in crowds along railway tracks decked out with triumphal arches in the red, white and blue of the British empire, to greet the arrival of the first seven carriages decorated with marigolds. On the platform a double file of mounted lancers, companies of turbaned sepoys, and the musicians of the gleaming Royal Brass Band were lined up. Alas, there was no radio or television in those days to immortalize the speeches exchanged by the representative of Victoria, Empress of India, and the sovereign who presided over this small corner of British India. 'I offer up a thousand thanks to the all-powerful God who has granted that Bhopal enjoy the signal protection of Her Imperial Majesty so that the brilliance of Western science may shine forth upon our land . . .' the Begum Shah Jahan had declared. In his response, the envoy from London

---

[1] Long garment worn by Muslim women, completely concealing the body and face.

extolled the political and commercial advantages that the railway would bring, not only to the small kingdom of Bhopal, but to the whole of central India. Then he raised his glass in a solemn toast to the success of the modern convenience, for which the enlightened sovereign had provided the funds. A firework display crowned the occasion. That day a piece of ancestral India had espoused itself to progress.

❊

For a long moment the Nadars hesitated without daring to take a step, so overwhelmed were they by the scene that greeted them as they got out of the railway carriage. The platform was packed with other landless peasants who had come there, like them, in search of work, and they found themselves trapped in a tide of people coming and going in all directions. Coolies trotted about with mountains of suitcases and parcels on their heads, vendors offered every conceivable merchandise for sale. Never before had they seen such sumptuousness: pyramids of oranges, sandals, combs, scissors, padlocks, glasses, bags; piles of shawls, saris, *dhotis*[2]; newspapers, all kinds of food and drink. Padmini and her family were bewildered, astounded, lost. Around them many of the other travellers appeared to be just as disorientated. Only Mangal the parrot seemed completely at ease. He never stopped warbling his joy and making the children laugh.

'Daddy, what are we going to do now?' asked the little girl, visibly at a loss.

'Where are we going to sleep tonight?' added her brother Gopal, who was holding the parrot cage above his head so

---

[2] Piece of material draped round the thighs and between the legs.

that his parents would see him if he got separated from them.

'We should look for a policeman,' advised the old man Prodip, who had been no more able than his son to decipher the work contract the *tharagar* for the railway had given them.

Outside the station, an officer in a white helmet was trying to channel the chaotic flow of traffic. Ratna cut a way through to him.

'We've just arrived from Orissa,' he murmured tentatively. 'Do you know if anyone from where we come from lives round here?'

The policeman signalled to him that he had not understood the question. It was hardly surprising. So many people speaking different languages got off the train at Bhopal.

Suddenly Padmini spotted a man selling samosas[3] on the far side of the square. With the sixth sense that Indians have for identifying a stranger's origin and caste, the little girl was convinced she had found a compatriot. She was not wrong.

'Don't worry, friends,' declared the man, 'there's an area occupied exclusively by people from our province. It's called the Orya Bustee[4] because the people who live there are all from Orissa like you and me, and speak Orya, our language.' He waved an arm in the direction of the minaret of a mosque opposite the station. 'Skirt that mosque,' he explained, 'and carry straight on. When you get to the railway line, turn right. You'll see a load of huts and sheds. That's Orya Bustee.'

Ratna Nadar bowed down to the ground to thank his

[3] Triangular-shaped fritters stuffed with vegetables or minced meat.
[4] Bustee: poor neigbourhood made up of makeshift shacks

25

benefactor by touching the man's sandals with his right hand, which he then placed on his head. Padmini rushed to the parrot's cage. 'We're saved!' she cried. The bird promptly responded with a triumphant squawk.

❄

As soon as he saw the little caravan approaching, the man seized his walking stick and went out to meet it. He was a hefty fellow of about fifty with a curly mop of hair and sideburns that joined up with the drooping ends of his moustache.

'Welcome, friends!' he began in a soft voice which belied his imposing appearance. 'I guess it's a roof that you're after here!'

'A roof would be a lot to hope for,' stammered Ratna Nadar apologetically, 'but perhaps just somewhere for me and my family to camp.'

'My name is Belram Mukkadam,' the stranger announced, pressing his hands together in front of his chest to greet the little group. 'I run the committee for mutual aid for neighbourhoods on the Kali Grounds.' He pointed in the direction of the string of sheds and huts on the edge of a vast empty expanse along the railway line. 'I'll show you where you can settle and build yourself a hut.'

Mukkadam was not an Adivasi, but he spoke the language of the people of Orissa. Thirty years earlier he had been the very first person to settle on the wasteland on the northern side of the city, bordering on what had once been the immense parade ground of the Victoria lancers, the cavalry regiment of the Nawabs of Bhopal. The hut he had built with the help of his wife Tulsabai and their son Pratap had been the first of the hundreds that now made up three neighbourhoods of improvised homes, in which several thousand immigrants from different Indian regions lived. Apart from the Orya bustee, there was the Chola bustee and the Jai Prakash bustee. Chola means 'chickpea'. It was by planting chickpeas on the edge of their encampments that the first occupants of the Chola bustee had escaped starvation. As for Jai Prakash, it was named after a famous disciple of Mahatma Gandhi who had taken up the cause of the country's poor.

His position as doyen of the three bustees had earned Belram Mukkadam a prerogative, never contested by the various godfathers of the local mafia who controlled those poor neighbourhoods where there was no municipal authority to intervene. He was the one who allocated newcomers the plots on which to make their homes.

Leading the Nadar family along a path that ran beside the railway track, he pointed to an empty space at the end of a row of huts.

'There's your bit of ground,' he said, tracing a square three yards by three yards in the dark earth with his tamarind stick. 'The committee for mutual aid will bring you materials, a *charpoy* and some utensils.'

Once more Ratna Nadar prostrated himself on the ground to thank this new benefactor. Then he turned to his family.

'The great god's anger is spent,' he declared. 'Our *chakra*[5] is turning again.'

❈

Orya Bustee, where the peasant from Mudilapa and his family had ended up, was the poorest of the three poverty-stricken neighbourhoods that had grown up along the parade ground.

---

[5] Chakra: wheel of destiny.

In the labyrinth of its alleyways, one sound singled itself out from all the others: that of coughing. Here tuberculosis was endemic.

There was no electricity. There was no drinking water, no drainage and not even the most rudimentary dispensary. There were scarcely even any traders, except for a travelling vegetable salesman and two small tea stalls. Very sweet milky tea sold in clay beakers was an important source of energy for many of the local residents. Apart from four skeletal cows and several mangy dogs, the only animals in evidence were goats. Their milk provided precious protein for their owners who, in winter, had no reservations about swaddling their animals in old rags, to prevent them from catching cold.

For all its poverty, Orya Bustee was unlike any of the other slums. Firstly, it had managed to maintain a rural feel, which contrasted with the jumble of huts made out of planks and sheet-metal in the other slums. Here all the dwellings were made out of bamboo and mud. These 'katcha[6] houses' were decorated with geometric designs drawn in rice paste to attract prosperity, just as they were in the villages of Orissa. They gave an area of detention-camp congestion an unexpected rural charm. Then there was the fact that the former peasants who had taken refuge there were not marginalized people. In their exile they had, after a fashion, reconstructed their village life. They had built a little temple out of bamboo and baked mud to house an image of Jagannath. Next to it, they had planted a sacred *tulsi*, a sort of arborescent basil with the power to repel reptiles, particularly cobras with their deadly venomous bite. The neighbourhood women were particularly devoted to their *tulsi* and a number of them made offerings to it, in order

---

[6] Katcha: crude earth.

that they might be cured of sterility. Here, as elsewhere in India, belief manifested itself in an uninterrupted succession of ritual festivities. A child's first tooth, the first haircut; a girl's first period, engagement, marriage, mourning; Diwali, the festival of lights, the Muslims' Eid and even Christmas . . . all of life's events, all festivals secular or religious, were publicly marked. For all their lack of education and despite their material poverty, the Adivasis of Orya Bustee had managed to maintain the rites and expressions of social and religious life that made up the rich and varied texture of their homeland.

# 4

# A *visionary billionaire to the rescue of humanity's food*

The crime committed by the infamous aphids in the Mudilapa fields would not go unpunished. All over the world armies of scientists and researchers were working relentlessly at destroying the miniature monsters. One of the chief temples dedicated to the crusade against the insects that plagued humanity's crops was an agronomical research centre in Yonkers, a residential suburb of New York on the banks of the Hudson river. It was called the Boyce Thompson Institute.

The man who founded this Institute was a billionaire with a messianic desire to commit his wealth to some great humanitarian cause. William Boyce Thompson (1869–1930) had amassed a huge fortune from copper mining in the mountains of Montana. In October 1917 the American Red Cross had made him a colonel and placed him in charge of an aid mission to Russia, then in the throes of revolution. The generous industrialist had swapped bow-tie and top hat for military uniform, and added a million dollars of his own money to the funds provided by the American government for the victims of the Russian famine. He came back from his journey to hell convinced that world peace depended on the equitable distribution of food, a conviction which was reinforced by his ardent faith in science and which led to the formulation of a spectacular philanthropic project. Because population growth was going to increase the need for food, it was vitally urgent to understand 'why and how plants grow, why they decline or flourish, how their diseases can be stemmed, how their development can be stimulated by better control of the elements that enable them to live'. The study of plants, so the generous patron claimed, could make a decisive contribution to humanity's well-being.

Out of this conviction was born, in 1924, the Boyce Thompson Institute for Plant Research, an ultramodern agronomical research centre, built on several acres of land less than an hour from New York City. Endowed by its founder with ten million dollars – a considerable sum at the time – the Institute incorporated chemistry and biology laboratories, experimental greenhouses and insect vivaria.

It was on the front line of the battle against plant-eating species that the Boyce Thompson Institute researchers achieved their first most significant victories, by eradicating the beetles that killed Californian pines and inventing subtle, sweet-smelling substances to lure the destructive little creatures into fatal traps.

At the beginning of the 1950s the *aphis fabae*, an aphid as devastating as the one which, a few years later, would obliterate the Mudilapa fodder crop, wrought havoc on the farmlands of the United States, Mexico, Central and South America. Found also in Malaysia, Japan and southern Europe, it attacked potatoes in addition to cereals, beetroot, fruit trees and vegetables, fodder and garden plants. This tiny predator has a beak equipped with two very fine piercing stylets, with which it sucks the sap from plants. As the Indian, Ratna Nadar, would so painfully discover, plants abruptly deprived of their vital substance wither and perish in days.

31

Before going in for the kill, this aphid, scarcely bigger than a pinhead, injects its victim with toxic saliva, causing hideous deformation of the stalks and leaves. To finish off the job, it exudes from its rectum honeydew to attract ants. These ants deposit a sort of soot on the leaves, which stifles any growth.

This was not the only nightmare parasite to afflict American and Asian peasants at that time. The red vine spider, an army worm that attacked food crops, and the striped stem borer joined forces with other destructive species to deprive humanity of a large part of its agricultural resources. Only the chemical industry could come up with a means of eradicating such a scourge. Conscious of all that was at stake, a number of companies went into action. One of them was American. Its name was Union Carbide.

❋

Born at the beginning of the century of a marriage of four companies that produced batteries and arc-lamps for acetylene street lighting and head-lamps for the first automobiles, 'Carbide' – as it was affectionately known by its staff – owed its first glorious hour to the 1914–18 war. It was helium from its stills that enabled tethered balloons to rise into the skies above France and spot German artillery fire; it was iron and zirconium-based armour-plating of its invention that thwarted the Kaiser's shells on the first Allied tanks; it was Carbide's active carbon granules in gas masks that protected the lungs of thousands of infantrymen in the trenches of the Somme and Champagne. Twenty-five years later, another world war was to enlist Carbide's services for America. Out of its collaboration with the scientists of the Manhattan Project, the first atomic bomb was born.

In less than a generation the absorption of dozens of

business concerns propelled the company to the forefront of the multinationals. By the second half of the century it was to feature amongst the mightiest of American companies, with 130 subsidiaries in some forty countries, approximately 500 production sites, and 120,000 employees. In 1976 it was to announce a turnover of six and a half billion dollars. The products that emerged from its laboratories, factories, pits and mines were innumerable. Carbide was the great provider of industrial gases such as nitrogen, oxygen, carbonic gas, methane, ethylene and propane used in the petrochemical industry as well as chemical substances like the ammonia and urea used in, among other things, the manufacture of fertilizers. It also produced sophisticated metallurgical specialities based on alloys of cobalt, chrome and tungsten, used in high-tensile equipment such as aeroplane turbines. Finally, it made a whole range of plastic goods for general use. Eight out of ten American housewives did their shopping with plastic bags stamped with the blue and white hexagon of Union Carbide. The logo also appeared on millions of plastic bottles, food packaging, photographic film and many other everyday items. The intercontinental telephone conversations of half the planet's inhabitants were conducted via underwater cables protected with sheathing made by Carbide. The antifreeze for one in every two cars, 60 per cent of batteries and silicone implants used in cosmetic surgery, the rubber for one in every five tyres, most aerosol fly and mosquito sprays, and even synthetic diamonds issued from the factories of a giant whose shares were amongst the safest investments on Wall Street.

From its imposing 52-storey aluminium and glass skyscraper at 270 Park Avenue, in the heart of Manhattan, Carbide determined the habits and dictated the choices of millions of men, women and children across continents. No other industrial company enjoyed the same degree of respectability – at least on the surface. After all, didn't people say that what was good for Carbide was good for America and therefore the world?

The production of pesticides was in line with its past and its experience. Its objective – ridding humanity of the insects that were stealing its food – could only enhance its international prestige.

# 5

## *Three zealots on the banks of the Hudson*

They looked more like Davis Cup players than laboratory researchers. Harry Haynes, aged thirty-four and Herbert Moorefield, aged thirty-six, both vigorous and fit-looking men, belonged to a profession that was relatively new. They were doctors of entomology. In July 1954, Union Carbide's management had rented an entire wing of the Boyce Thompson Institute in Yonkers for these two eminent experts. It had further strengthened the team by adding to it one of the most brilliant staff members at its South Charleston research centre, the 38-year-old chemist, Joseph Lambrech. To these three exceptionally gifted people the company entrusted a mission of the utmost importance: that of devising a product to exterminate a wide range of parasites whilst respecting prevailing standards for the protection and safety of humans and the environment. That summer, on the top floor of the multinational's New York head office, no one was in any doubt: the company that managed to reconcile these two objectives would walk away with the world pesticide market.

Lambrech gave the object of his labours a code name. For

convenience' sake the 'Experimental Insecticide Seven Seven' would soon become 'Sevin'.

Going through all his predecessors' studies with a fine-tooth comb, the chemist slaved away at combining new molecules, hoping to find one that would kill aphids, red spiders and army worms without leaving too many toxic residues in the vegetation and environment. For months on end, his entomologist colleagues tested his combinations on leaves, stems and ears infested with all kinds of insects. In its hundreds of cages and containers, the Boyce Thompson Institute harboured an unimaginably rich zoo of the infinitely small. It also had acres of glass houses in which all the climates of the planet could be recreated around a limitless variety of plants. In large glass cases the different molecules could be tested with sprays of varying doses, directed from every possible angle at samples of every variety of crop. The entomologists, Haynes and Moorefield, would then deposit colonies of larvae, caterpillars and other insects raised in their laboratories on the treated surfaces. Hour by hour they would watch over their subjects' agony. They collected up the corpses on glass slides, examined them under the microscope, and subjected the plants and soil to detailed analysis to find any traces of chemical pollution. Their observations would enable their chemist colleague to hone the production of an insecticide ever nearer to what was required.

After three years of intense effort the team came up with a combination of a methyl derivative of carbamic acid and alpha naphtol, in the form of whitish crystals soluble in water. Those three years had been taken up with hundreds of experiments, not just on all known species of insects, but also on thousands of male and female rats, on rabbits, pigeons, fish, bees and even prawns and lobsters. Finally, one evening in July 1957, the three zealots in Yonkers together with their wives, were able to crack open the champagne. The god DDT might have had to be eliminated, but agriculture was not to remain defenceless. Sevin, born on the banks of the Hudson, would soon put a weapon in the hands of all the world's peasants.

❅

Carbide was quick to flood America with brochures proclaiming the birth of its miracle product. There was no end to the praises sung to it, and to underline its total lack of toxicity, photographs depicted Herbert Moorefield, one of the inventors of Sevin, in the process of tasting a few granules with all the alacrity of a child licking his chocolate-coated lips. According to the publicity, Sevin protected an infinite range of crops: cotton, vegetables, lemons, bananas, pineapples, olives, cocoa, coffee, sunflowers, sorghum, sugar cane and rice. You could spread it on maize, alfa-grass, beans, peanuts and soya right up until harvest time, with no danger of any toxic residues. It worked just as well on adult insects as it did on eggs and larvae. It was so effective that it even poisoned parasites that had become resistant to other insecticides. Its potency was not limited to agricultural crops. A few ounces of Sevin spread round the outside of homes or sprayed on the walls, frames or roofs exterminated mosquitoes, cockroaches, and other bugs harmful to family life. Better still, Sevin controlled the number of fleas, lice and ticks on dogs, cats and farmyard animals, without putting their lives at risk. In short, Sevin was precisely the magic panacea the multinational's new agricultural products division had been waiting for to send its turnover sky high.

No one was more convinced of this fact than a young, 29-year-old Argentinian agronomical engineer. Handsome and charming, Eduardo Muñoz came from a well-to-do

Buenos Aires family. He had chosen agronomy as an act of defiance after his failure to pass the entrance examination for the diplomatic service. He had married an attractive American girl who worked at the United States embassy, and so found amongst his wedding presents the perfect incentive to set off for pastures new, the celebrated 'green card', a permit to work in the land of Uncle Sam. Out of the fifty offers he received when he sent off his curriculum vitae, he chose the first. It came from Union Carbide. A year's training on the various company sites and a monthly salary of 485 dollars had turned the handsome Argentinian into a proper Carbider. The invention of Sevin was to provide him with the opportunity to exercise his extraordinary talents as a salesman. In Mexico, Columbia, Peru, Argentina, Chile, Brazil . . . soon there was not one single grower who was not aware of the merits of the American pesticide. At agricultural fairs, harvest competitions, farmers' meetings, Muñoz was everywhere with his flags to the glory of Sevin, his on site demonstrations, his handouts, and his sponsored raffles. It was almost inevitable that Central and South America would one day become too restrictive for the indefatigable travelling salesman. He would have to find other places in which to satisfy his passion for selling.

# 6

## *The daily heroism of the people of the bustees*

'Here, brother, it works out cheaper to sweat a fellow to death than hire a buffalo,' remarked Belram Mukkadam to Padmini's father, who had just come back from a day's work on the railway line.

The sturdy date palm climber from Orissa was reeling with exhaustion. All day long he had carted sleepers and heavy steel tracks about. The coolies the railway management had recruited were all immigrants like him, forced into exile by the poverty of the countryside.

From the outset, this slave labour had been terrible. Ratna Nadar felt himself growing weaker by the day, stricken with nausea, cramps, bouts of sweating, and dizziness. His muscles wasted visibly. Soon he had difficulty in remaining upright. He suffered from hallucinations and nightmares. He was the victim of what specialists call 'convict syndrome'. The small quantity of rice, lentils and occasional fish that he bought before leaving for work in the morning was for him. It is a tradition amongst India's poor that the family food be kept for the rice earner. Even so, often the lack of fuel prevented Sheela from cooking her husband's food. It took several weeks before Ratna felt his strength returning. Only then could the whole family eat almost its fill.

For Padmini and her brother Gopal, the brutal immersion in the overpopulated world of city workers was just as painful. Every day they saw sights that shocked the sensibilities of children raised in the countryside.

'Gopal, look!' cried Padmini one morning, pointing to a gang of youngsters scaling the back of a stationary locomotive.

'They're out to pinch bits of coal,' Gopal explained calmly.

'They're thieves!' Padmini was indignant, furious that her brother did not share her outrage.

Her eyes filled with tears.

'Dry your eyes, little one! You too will steal coal to make *ladhus*[1]. If you don't your mother won't be able to cook anything for you to eat.'

The man who had just spoken had no fingers left on his right hand. Padmini and her parents would come to know and respect this prominent figure in Orya Bustee. At thirty-eight, Ganga Ram was a survivor of leprosy, a curse which still afflicts 5 million Indians today. Thrown out on the streets by the owner of the garage in Bombay where he used to wash cars, Ram had ended up in the communal ward of Hamidia Hospital in Bhopal. He had been treated and cured, and had a certificate given to him by a doctor to prove it. Uncertain where to go and what to do, for seven years he had remained in the wing for contagious diseases, performing small services for the sick and their nurses. He had applied dressings, changed the incontinent, administered enemas and even given injections. One day, he was called upon to transport an attractive woman of about thirty with luminous green eyes. A truck had broken both her legs. Her name

---

[1] Small balls of coal and straw used as fuel for cooking food.

was Dalima, and it was love at first sight. In the communal ward Dalima had adopted a 10-year-old orphan who had been found half dead on a pavement. He had been taken to hospital in a police van. His name was Dilip. Lively and alert, this skinny urchin with short-cropped hair, who was always ready to help, was the darling of the occupants of the communal ward. A few weeks later the former leper, Dalima and young Dilip left the hospital to settle in Orya Bustee where, with the tip of his walking stick, Belram Mukkadam had assigned them a place on which to build a hut. Some of the neighbours brought bamboo canes, planks and a piece of canvas, others brought cooking utensils, a *charpoy* and a little linen. 'All we had by way of luggage were Dalima's crutches,' Ganga Ram said.

For months they survived on Dilip's resourcefulness alone. He was the one who inveigled the neighbourhood children into stealing bits of coal from the railway engines. One morning, he persuaded Padmini to go with him.

'You have to hurry up, little sister. The railway police are on the look-out.'

'Are they nasty?' The little girl was worried.

'Nasty!' The boy burst out laughing. 'If they catch you, be prepared to give them a fat backsheesh. Otherwise they'll take you away in a van and there' – Dilip made a gesture, the meaning of which eluded the little peasant girl.

When they got back from their expedition, the slum midwife, the elderly Prema Bai, who lived in the hut opposite the Nadars, gave her young neighbour a little straw and some nanny-goat droppings.

'Crumble the coal with the straw and the droppings and knead the whole lot together for a good while,' she instructed, 'then make little balls out of it and put them to dry.'

An hour later Padmini took the fruits of her crime triumphantly to her mother.

'Here you are, mother, *ladhus*. Now you'll be able to cook father's food.'

✳

For peasants used to the sovereign silence of the country-side, the infernal din of the trains passing in front of their huts was a great trial. Their lives revolved round the rhythm of the incessant coming and going of dozens of trains. 'I got to know their timetable, to know whether they were on time or late,' Padmini recalled. 'Some of them, like the Mangala Express, made our huts shake as they roared past in the middle of the night. That was the worst one. The Shatabdi Express to Delhi went by in the early afternoon and the Jammu Mail just before sunset. The drivers must have had fun, terrifying us with the roar of their engines.'

There were some advantages to being so close to the railway tracks. When a red light brought a train to a halt outside the huts, the drivers would throw a few coins for the children to run and buy them some *pan*[2]. There was always some small change left over.

'Watch where you put your feet when you're walking between the rails,' Dilip advised Padmini. 'That's where people come to do the necessary.'

Fortunately, the tracks were also strewn with a multitude of small treasures that people on the trains had thrown away. There were bottles, old tubes of toothpaste, flat batteries, empty tins, plastic soles, shreds of clothing, and pieces of rag to be picked up. Dilip used to negotiate a price for them with a ragpicker who came round every week. The daily takings could be as much as three or four rupees. On Magnet cigarette packets there was an illustration of the Taj Mahal.

---

[2] Quid of betel-nut for chewing.

Dilip and Gopal, Padmini's brother, would cut out the picture to make playing cards, which they sold on the station platforms. 'I shall never forget the Orya Bustee trains,' Padmini said. 'They brought a little excitement and joy into our difficult life.'

One of these joys came from an unexpected source. Every morning, Padmini's mother and her neighbours would take up their positions along the railway line to wait for the arrival of the Punjab Express. On their heads they carried buckets, bowls and basins. As soon as the train stopped, they would rush to the engine.

'May the great god bless you!' they would call out in a chorus to the driver. 'Will you turn on your tap for us?'

If he was kindly disposed, the driver would undo the valve on his boiler and fill their containers with a few gallons of a commodity to which few of Bhopal's poor had access: hot water.

# 7

## An American valley that
## ruled the world

Dilip had an eye for these things. He saw at once that the hut built by Ratna Nadar and his family would never survive the onslaughts of the monsoon.

'You should double the roof supports,' he advised Padmini.

The little girl gave a gesture of helplessness.

'We haven't even the money to buy incense sticks for the God,' she sighed. 'It's three days since grandpa and grandma have eaten. They refuse to sacrifice the parrot.'

Dilip took a five rupee note out of his shorts.

'There,' he said, 'that's an advance on our next treasure hunt along the railway track. Your father will be able to buy two bamboo poles.'

❄

On the other side of the world, in a lush valley in West Virginia, a team of Union Carbide engineers and workmen were putting in the girders for a new factory destined to be the multinational's flagship. The Kanawha Valley had long served as a fief of the company. Curiously enough it owed its other name, 'magic valley', to the most ordinary of resources: its salt beds. With reserves of up to almost a billion tons, the valley had attracted people and animals since prehistoric times. Salt had made wild animals carve pathways through the forest to the saline pools along the river. It had sent Indians along the same routes in pursuit of game, then provided them with the brine in which to preserve their kill. In the seventeenth century it had drawn a few intrepid explorers to this inhospitable region. And white gold was not the magic valley's only trump. The primitive forests that covered it had provided the material necessary to build houses, boats, barges, butts in which to transport the salt, carts, bridges and mill wheels. An entire wood industry had grown up along the Kanawha. Connecting directly with the Ohio and Mississippi rivers, the river gave the valley's merchandise and travellers access to the centre and the south of North America.

At the beginning of the First World War, the valley was also found to contain prodigious energy resources. The discovery of oil, coal and natural gas had precipitated the Kanawha into the world of the chemical industry with its apparently limitless horizons. The 1920s had seen the region's woodlands replaced by other forests, metal ones, with chimneys, towers, flares, reservoirs, platforms, and pipe and tube work. These new factories belonged to giants like Dupont de Nemours, Monsanto and Union Carbide. It was there, on its Institute site, and in its research centre a few miles away from the peaceful little town of Charleston, that Carbide's chemical engineers had come up with the innumerable innovative products that were to transform the lives of billions. Turning chemistry into the Mr Fix-it of everyday life, they had helped to revolutionize products as varied as fertilizers, medicines, textiles, detergents, paints, photographic films . . . The list was endless but, because the chemical industry is not quite like any other, the revolution had

45

its price. Many of the substances that produce these everyday goods are as dangerous as the radiation produced by the nuclear industry. Ethylene oxide, involved in the manufacture of ordinary vehicle antifreeze, is as deadly as plutonium dust. Phosgene, one of the components currently most used, asphyxiated thousands of First World War soldiers in the guise of 'mustard gas'. Hydrogen cyanide, a gas with a pleasant almond smell used in medicines, was adopted by a number of American prisons to execute those condemned to death. In its Kanawha Valley factories, Carbide alone produced 200 chemical substances, many of which, like chloroform, ethylene oxide, acrylonitrile, benzene and vinyl chloride, were known for their tendency to cause cancers in humans and in animals.

Like its competitors, the Carbide company tried to keep its reputation intact by devoting substantial sums to ensuring the safety of its staff in the workplace and to a strict policy of safeguarding the environment. All too often firms awarded themselves certificates of good conduct, even as their toxic waste was insidiously poisoning the lush Kanawha countryside. Though these self-congratulatory measures were widely reported by complacent members of the media, they did not always achieve their desired end. Carbide was to find itself condemned to paying heavy fines for having poured highly carcinogenic products into the Kanawha River and the atmosphere. An enquiry conducted in the 1970s was to reveal that the number of cancers diagnosed in the occupants of the valley was 21 per cent higher than the American national average. The incidences of cancers of the lung and endocrine glands, and leukaemia in particular, were among the highest in the country. One study carried out by the state of West Virginia's health department found that people living in areas down wind of the South Charleston and Institute factories presented twice as many cancerous tumours as the rest of the population of the United States. These findings would not, however, prevent Carbide from constructing on its Institute site a completely innovatory factory, to manufacture Sevin, the beacon insecticide the company wanted to distribute throughout the world.

This high-tech project modified the procedure that the three researchers at the Boyce Thompson Institute had used to invent Sevin. It introduced a chemical process that would both substantially reduce production costs and eliminate waste. The manufacturing process involved making phosgene gas react with another gas called monomethylamine. The reaction of these two gases produced a new molecule, methyl isocyanate. In a second stage the methyl isocyanate was combined with alpha naphtol to produce Sevin. More commonly known by its three initials, MIC, methyl isocyanate is one of the most dangerous compounds ever conceived by the sorcerer's apprentices of the chemical industry. Carbide's toxicologists had tested it on rats. The results were so terrifying that the company banned publication of their work. Other experiments had shown that animals exposed to MIC vapours alone died almost instantaneously. Once inhaled, MIC destroyed the respiratory system with lightning speed, caused irreversible blindness and burnt the pigment of the skin.

German toxicologists had dared to go further by subjecting voluntary human guinea pigs to minute doses of MIC. Although disapproved of by the scientific community, these experiments did make it possible to determine the threshold of tolerance of exposure to MIC, in the same way that the level of tolerance to nuclear radiation had been established. The research was all the more helpful because thousands of workers all over the world, making synthetic foam products, such as insulation panelling, mattresses and car seats, found themselves in daily contact with other isocyanates,

47

cousins of MIC. Thanks to its new factory, Carbide would be able to sell MIC to all those manufacturers who used iso-cyanates, but who were reluctant to take on the dangers involved in their production. Most importantly of all, the American company was going to be able to sell Sevin all over the world.

# 8

## A little mouse under the seats of Bhopal's trains

*T*HE BHOPAL TEA HOUSE. There was something faintly comical about the sign. Its faded letters were displayed across the façade of a booth made out of planks that stood almost opposite the entrance to Orya Bustee. There, amidst the nauseating smell of frying fat, the traditional very sweet tea with milk, millet flour fritters, minced chillies and onions, rice and dhal, chapattis and other kinds of griddle cakes were served. Its main trade, however, was in 'country liquor', a local gut rot made out of fermented animal intestines, of which the tea house sold gallons every day. A notice in English warned clients that the establishment did not give credit: *YOU EAT, YOU DRINK, YOU PAY, YOU GO.* The proprietor, a pot-bellied Sikh with bushy eyebrows, rarely showed himself. Although one of the most important local figures, Pulpul Singh made his presence felt elsewhere. He was the money lender for the three bustees, a profession which he exercised from behind the heavy grilles of his two-storey modern house at the entrance to Chola. Enthroned like a Buddha in front of his Godrej-stamped safe and two immense chromos of the Golden Temple of Amritsar and a portrait of Guru Nanak, the venerable founder of the Sikh community, Pulpul Singh exploited the economic misfortunes of the poor. To recover his debts, he had engaged a convict on the run from a Punjabi prison. With a filthy turban on his head and his dagger ever at the ready, he had the protection of the police whom he bribed on behalf of his master. This villain was the terror of small borrowers. So hated was he that his master had stopped entrusting the management of his drinks stall to him. Instead he used the man most respected by the local people, Belram Mukkadam, whose walking stick had marked out the site for all the residents' huts.

Founder of the Committee for Mutual Aid, which combatted injustice and fought to relieve the worst cases of distress, Mukkadam was a legend in his own lifetime. For thirty years he had battled ceaselessly with corrupt officials, shady politicians, property agents and all those who wanted to get rid of the ghettos on the belt of land north of town. Because of him the date, 18 August 1978, would become famous in the history of Bhopal. On that day 2000 poverty-stricken people led by Mukkadam invaded the local parliament to demand the cancellation of an eviction operation planned for the next day. He would force the poor to hold their heads up high, strengthen their spirit of resistance, and gather about him men, united regardless of religion, caste or background, who formed a sort of hidden government for the bustees.

Despite the fact that a yawning divide separated this hero from the sordid activities of his employer, Mukkadam had agreed to take on the management of the Bhopal Tea House because it would provide him with a forum. Around its handful of tables reeking of alcohol, people could publicly discuss their affairs and better organize their response to any imminent danger.

50

The little girl bounded towards the dishevelled looking man who had just appeared at the end of the alleyway, staggering about like a drunk.

'Daddy, Daddy!' she cried as she ran towards her father.

Clearly he had stopped at Belram Mukkadam's tea house. He was not a drinker but Padmini's father had downed a few glasses of country liquor. It was an indication that something serious had happened. Padmini threw herself at his feet.

'The railway work is finished,' mumbled Ratna Nadar with difficulty. 'They've thrown us out.'

On that winter's day more than 300 coolies had suffered the same fate. There were no employment laws to protect temporary workers. They could be laid off at any time without notice or indemnity. For the Nadars, as for all the other families, it was a terrible blow. 'My father tried desperately to find another job,' Padmini said. 'Every morning, he would set off in the direction of Berasia Road in the hope of meeting a *tharagar* who would take him on for a few hours or a few days to pull carts or carry materials. But there wasn't any building work going on that winter in the area. Once again our stomachs began to rumble.'

One evening when the whole family was preparing to go to bed without food, Sheela decided to surprise them. She lined up all their bowls on the beaten earth floor and filled them with a glutinous gruel generously sprinkled with aromatic curry powder.

'Be careful not to swallow the little bones,' she cautioned.

They all understood what she was saying. She had cooked the parrot.

The next morning, Padmini saw Dilip in the doorway to her lodging.

'Come with me, and I promise you no one in your hut will ever go hungry again,' he declared with authority.

The little girl surveyed the boy's torn clothes with concern. His shorts and shirt were spotted with bloodstains.

'Where do you want to take me?' she asked, worried.

Dilip pointed to the amulet he wore round his neck.

'Don't be frightened. With this we won't be in any danger.'

They walked along the railway line in the direction of the station. On the way, Dilip stopped at a pile of garbage and began to scratch furiously at it.

'Look, Padmini!' he exclaimed, brandishing two small brushes he had just unearthed. 'These'll earn you lots of rupees.'

At the station Dilip met up with the members of his gang.

'Hi there, boss!' called out one of the urchins, who was also armed with a small brush.

'No luck, the Delhi train's late,' announced another boy.

'And the one from Bombay?' asked Dilip.

'Not announced yet,' replied a third who was wearing a little Muslim garrison-cap on his head. The gang members belonged to all different faiths.

Dilip introduced Padmini to his companions, who nodded their heads in admiration.

'With such a pretty mouse, we're bound to make a fortune!' laughed the eldest.

The sound of a whistle cut short their conversation and galvanized the small group into activity. Dilip dragged Padmini along by the hand onto the other platform. The man who had blown the whistle was an inspector with the railway police. He and another policeman were about to launch themselves after the gang when Dilip raised his arm.

'I'm coming!' he called out.

Clambering over the rails with feline agility, he joined the policemen. Padmini saw her friend take a note out of the pocket of his shorts and slip it discreetly into the inspector's hand. It was standard practice. Then the Delhi train arrived. The gang members spread themselves along the platform so that the various carriages were divided up between them. Dilip pushed Padmini towards the first open door. He pointed out to her the rows of seats onto which the passengers were piling.

'Get down on all fours, crawl along with your brush and pick up anything you can find,' he told her. 'But hurry up! We have to get off at the next stop to come back to Bhopal!'

Padmini sneaked under the first row of seats, sweeping as frantically as if she were prospecting for gold. Suddenly, between the feet of one of the passengers, she noticed a piece of chapatti. 'I was so hungry I lunged at it and swallowed it,' she admitted. 'Luckily people had also thrown away some banana skins and orange peel.' The little sweeper quickly gathered all this and more. At the first stop, the gang made an inventory of their findings.

'Guess what I've got in my hands,' she cried, holding her closed palm in front of the boy's eyes.

'A diamond the size of a bottle stopper!'

'Idiot!' laughed Padmini and opened her hand to reveal two small five paisa coins, 'I'll be able to buy my father two *bidis*.'

'Well done!' said Dilip with obvious excitement. He took from his waist a sock, a used battery, a sandal and a cone of newspaper full of peanuts. 'I'll sell all this to my usual ragpicker. He should give me three or four rupees.'

That evening Dalima's son brought his young accomplice a ten rupee note. He had generously rounded up the sum he had received from the ragpicker. Padmini caressed the note for a long moment. Then she sighed:

'We're saved.'

Padmini came to have her favourite trains and know all their conductors. Some of them would give her one or two rupees and sometimes a biscuit when they came across her in the act of sweeping under the seats or in the corridors. But there were also the *big dadas*[1] in Bhopal station. Always out for a fight, they would try and take whatever the sweepers had collected. They were in cahoots with the police and if Dilip did not give them ten or twenty rupees, out came the clubs.

'Often they would manage to snatch our entire day's takings,' Padmini said. 'Then I would go home empty-handed and my mother and brother Gopal would start crying. Sometimes when the trains were running late, I would spend the night with Dilip and his gang in the station. When it was very cold, Dilip would light a fire on the platform. We would lie down next to the flames to sleep until the next train came through. There were times too when we slept in other stations, at Nagpur, Itarsi or Indore, waiting for a train to take us back to Bhopal.'

It was in one of these stations that one night Dilip and his companions would lose their little Adivasi sister.

---

[1] 'Big arms', ruffians.

# 9

## A poison that smelt like boiled cabbage

'Fatal if inhaled!' Displayed on labels marked with a skull and crossbones, posters and printed pages in user manuals, the warning was directed at the manufacturers, transporters and users of MIC. The molecule had such a volatile nature that it had only to come into contact with a few drops of water or a few ounces of metal dust for it to break into an uncontrollably violent reaction. No safety system, no matter how sophisticated, would then be able to stop it emitting a fatal cloud into the atmosphere. To prevent explosion, MIC had to be kept permanently at a temperature near zero. Provision must therefore be made for the refrigeration of any drums or tanks that were to hold it. Any plant that was going to carry stocks of it must, moreover, be equipped with decontamination apparatus and flares to neutralize or burn it in case of accidental leakage. The transportation of methyl isocyanate was subject to extraordinary safety precautions. Lorry drivers were required to 'avoid congested routes, bypass towns and cities, and stop as infrequently as possible'. In case of a sudden burning sensation in the eyes, they were to rush to the nearest telephone box and dial the *HELP* number, followed by 744 34 85,

Carbide's emergency number. They must then evacuate their vehicle to 'an unoccupied area'.

Carbide had decided to play its hand openly, which was not always the case in the chemical industry. A whole chapter of its manual detailed the horrible effects of accidentally inhaling MIC: first severe pains in the chest, then suffocation and, finally, pulmonary oedema and possible death. In case of such an incident the thorough rinsing of any contaminated parts with water, the liberal use of oxygen and the administration of medication to dilate the bronchia were the recommended measures.

All the same, Carbide did not disclose all the information revealed by two secret studies undertaken at its request in 1963 and 1970 by the Mellon Institute of the Carnegie Mellon University in Pittsburgh. These studies of the toxicity of methyl isocyanate showed that under the influence of heat it broke down into several molecules, which were also potentially fatal. Among these molecules was hydrocyanide acid, a gas with a sinister reputation, which when inhaled in strong doses almost invariably caused immediate death. The two studies also revealed, however, the existence of an effective antidote to this fatal gas. Injection with sodium thiosulphate could, in certain cases, neutralize the deadly effects of hydrogen cyanide. Carbide had not seen fit to include this information in its documentation for MIC.

※

It was in its new Institute plant, on the banks of the Kanawha, that Carbide intended making the MIC it needed for its annual production of 30,000 tons of Sevin. Known as Institute 2, this plant was to operate in conditions so safe and with such regard for the environment that it would be an industrial model for the entire valley. Anchored in a sea of concrete, its metal structures were spread over five levels. Each was crammed with reactors, distillation columns, tanks, flares, condensers, furnaces, exchangers, pumps and a network of dozens of miles of piping of varying sizes and colours, according to what liquid or gas it conveyed.

'It was a really beautiful plant,' the American engineer, Warren Woomer recounted. He had joined Carbide at the age of twenty-two and had become an expert on high risk plants. 'It's true that you had a sense of danger when you went in there. But I had got used to living among toxic substances. After all, chemical engineers spend their lives in contact with dangerous products. You have to learn to respect them and, above all, you have to get to know them and learn how to handle them. If you make a mistake, there's very little chance they'll forgive you.'

Warren Woomer knew that the piloting of this 'high tech' factory had been entrusted to the best professionals in the field. To belong to the MIC production unit was considered an honour on the Institute site. It also had its advantages: salaries there took into account the hazardous nature of the substances used. They were the highest in the company.

Carbide had provided the plant with an impressive arsenal of security systems. There were countless decontamination towers and flares capable of neutralizing and burning off large quantities of gas in case of accidental leakage. Hundreds of valves meant that any fluid showing an abnormal pressure could be evacuated into diversion circuits. Successions of thermostatic sluice-gates, one way valves, joints, rupture discs, temperature sensors and pressure gauges watched over all the sensitive equipment and over the piping which had itself been put together with high resistance welding and checked by X-ray. Damping devices prevented any excessive movement of the metal. As in the most modern airplanes, the electric circuitry had been duplicated and protected to resist the onslaught of even the most corrosive acids. In the event of electricity failure, super-powerful generators would immediately cut in. Special double-skinned piping had been installed to conduct the MIC to its storage tanks. Between the skins a flux of nitrogen was circulated. Every ten yards sensors checked the purity of the gas. The tiniest escape of MIC into the nitrogen would be detected immediately and trigger an alarm and immediate intervention.

To ensure total reliability, the builders of Institute 2 had designed high performance equipment, produced by companies like International Nickel and Ingersol Rand, among the United States' most eminent specialists in alloying and mechanical engineering.

No less exceptional precautions had been taken to ensure the safety of the staff. A network of loudspeakers and sirens, modulated differently according to the nature of the incident, was ready to go into action at the slightest alert. Crews of firemen specializing in chemical fires and a system of automatic sprinklers could flood the factory with carbonic foam in a matter of minutes. Dozens of red-painted boxes on every level equipped the workers with protective suits, breathing apparatus, ocular rinses and decontamination showers. The plant was even equipped with a monitoring system that was constantly analysing samples taken from the atmosphere. If the safety level was exceeded, a loud alarm would sound and the location of the anomaly would appear on a screen.

With its walls studded with pressure gauges, levers and buttons the control room looked like the flight deck on Concorde. Day and night, different coloured markers traced the plant's every breath on rolls of graph paper. Day and night, keys, levers and handles relayed electronic orders to open or close the stopcocks, shut down or activate a circuit, launch or interrupt a production or maintenance operation. One of the dials most carefully monitored was a temperature gauge. It was linked to thermometers located on each of the tanks of methyl isocyanate used in the continuous production of Sevin. Given that the needles on these instruments must never rise above zero degrees, the builders of the American factory had lined the walls of the tanks with a skein of coils that circulated cooling chloroform.

❋

It was on the smell, or rather the lack of it, that the initial results of these unprecedented efforts were judged. A properly sealed chemical plant does not give off any smell. That was not the case with the other factories polluting the Kanawha Valley with emissions that none of its 250,000 local residents could escape. 'The smells ended up permeating the trees, flowers, the river water and even the air we breathed,' complained Pamela Nixon, laboratory assistant at the Saint Francis Hospital in Charleston, who lived with her family, along with several hundred other black families on modest incomes, in the Perkins Avenue area, close by tanks and chimneys of the Institute works. A few days before the launch of the new factory, Pamela and her neighbours found a leaflet in their mailboxes, sent by Union Carbide's local management. Entitled *Plan for the general evacuation of Institute*, this document listed the procedures to be observed in case of incident. The first instruction was not to try to get away. 'Switch your radio to WCAW station, 689 metres medium wave, or your television to channel 8 on station WCHS,' the document instructed. 'This is the kind of announcement that you are likely to hear: *At ten o'clock this morning, the West Virginia State police reported an industrial accident involving dangerous chemical substances.*

*The accident occurred at 09.50 hours at the Institute site of the Union Carbide Company. All persons living in the vicinity are invited to remain in their homes, close their doors and windows, turn off all fans and air-conditioning systems, and keep a listening watch for further instructions. The next communication will be broadcast in five minutes.'*

Pamela Nixon sellotaped the sheet of paper to a corner of her fridge door.

Two weeks later, when the new plant had begun normal production, the young woman suddenly noticed a strange smell coming in through her kitchen window. It was being carried on the breeze blowing, as usual, from the direction of the industrial structures located upwind of her home. It was not the smell of fish or rotten eggs that she was used to from the other factories in the valley. This new smell went to show that even if the plant she could see from her house was a model of advanced technology, it was not, in fact, totally sealed. Casting her mind back, Pamela recalled a childhood memory. Like her mother's cooking every Sunday after church, the methyl isocyanate produced by the Union Carbide Company smelled like boiled cabbage.[1]

---

[1] This smell of boiled cabbage was to take hold in the magic valley. The Federal Agency for the protection of the environment (EPA) revealed that, between 1 January 1980 and the end of 1984, sixty-seven leakages of methyl isocyanate occurred at the Institute factory. The management of the factory took care not to bring these leaks to the attention of the people living in the valley, considering that none of them had posed a real threat to health, or exceeded the legally accepted standards for toxic emissions in the atmosphere.

# 10

## *They deserved the mercy of God*

The unusual figure that came into view one morning, at the entrance to Orya Bustee, took Belram Mukkadam by surprise. He had never before seen a European venture into the neighbourhood. Tall, dressed in a black, ankle-length robe, with a metal cross strung across her chest, her grey hair boyishly cropped and thick round glasses taking up much of her thin face, she sported a luminous smile. Mukkadam welcomed her with his customary friendliness.

'What a pleasant surprise! Welcome, Sister. What wind of good fortune brings you here?' he asked.

The visitor greeted him Indian-style.

'I've heard your neighbourhood needs someone to provide medical care for the sick, the children and the elderly. Well, here I am. I've come to offer you my humble services.'

Mukkadam bowed almost to the ground.

'Bless you, Sister! God has sent you. There's so much suffering to be relieved here.'

Sister Felicity McIntyre was Scottish. Born into a diplomatic family that had spent long periods in France, at eighteen she had entered a missionary order. Sent first to Senegal, then to Ceylon and finally to India, for fourteen years she had been living in Bhopal where she ran a centre set up by the diocese to take in abandoned children. Most of them were suffering from serious mental handicap. The centre had been established in a modern building in the south of the city. It bore the beautiful name of 'Ashinitekan – House of Hope'. Above the entrance the nun had nailed a plate with the inscription: 'When God closes one door, he opens another'. Down's Syndrome children; autistic children; those with tuberculosis of the bone or polio; blind, deaf and mute children – all lived together in a single large room with pale green walls, decorated with pictures of Mahatma Gandhi and Jesus Christ.

The room was lively with activity. Several young girls trained by Sister Felicity busied themselves with the children, coaxing them to move, perhaps walk or play. Parallel bars, rubber balls, swivel boards and small pedal-cars took the place of physiotherapy equipment. Here life was stronger than any misfortune. Many of the patients needed special care. They had to be dressed, fed, taken to the toilet, washed. Above all, their intelligence had to be awakened, something which de-manded endless patience and love. Sister Felicity shared her bedroom with a mentally retarded 12-year-old who seemed more like six. Suffering from cerebral palsy, Nadia was as dependent as a baby. But her smile proclaimed her will to live and her gratitude. Although she refuted the idea, Sister Felicity was to Bhopal what Mother Teresa was to Calcutta.

Mukkadam led the nun through the labyrinth of alleyways.

'This is a really wretched place,' he apologized.

'I'm used to it,' his visitor reassured him, greeting those who gathered along her way with a cheerful 'Namaste[1]'.

---

[1] Namaste or Namaskar, literally 'to prostrate oneself'. A salutation involving pressing the hands together level with heart or face. The degree of respect shown is measured by the height at which the hands, which may be raised as high as the forehead, are held.

She went into several huts and examined some of the children. Rickets, alopecia, intestinal infections . . . Orya Bustee had the full collection of diseases found in poverty-stricken neighbourhoods. The nun was on familiar ground and no stranger to the slums. She had learned to receive the confidences of the dying, to watch over the dead, to pray with their families, wash corpses and accompany the deceased on their last journey to the cemetery or the funeral pyre. Above all, with the assistance of her large, black bag full of medicines, vials of serum, antibiotics, morphine and even small surgical instruments, she had treated people, comforted and cured them.

'I'll come every Monday morning,' she announced in Hindi. 'I'll need some families to take turns at letting me use their huts.'

The suggestion gave rise to an immediate commotion. All the mothers were prepared to offer the white *didi*[2] the use of their lodgings so that she could care for the occupants of the bustee.

'And then I'll need a volunteer to help me,' she added, casting a discerning eye round the faces crowded about her.

'Me, me, *didi*!'

Felicity turned to see a little girl with slanting eyes, standing in front of the hut.

'What's your name?'

'Padmini.'

'All right, Padmini, I'll take you on trial as my assistant in our small clinic.'

❊

---

[2] Big sister

On the following Monday an expectant line had formed in the alleyway in front of Padmini's house, well before Sister Felicity arrived. Padmini had tried to sort out the most serious cases in order to take them first into the hut that her parents had turned into an improvised infirmary. More often than not, they were rickety babies with swollen stomachs whom their mothers held out to the nun with a look of entreaty. 'In all my years of working in Africa, Ceylon and India, I had never seen such cases of deficiency diseases. The fontanelles had not even closed up in many of the children. The bone of their skulls had become deformed for lack of calcium and their dolichocephalic features made them look a bit like Egyptian mummies,' Sister Felicity recounted.

Tuberculosis might be the number one killer in Orya Bustee and its neighbouring slums, but typhoid, tetanus, malaria, polio, gastro-intestinal infections and skin diseases caused damage that was often irreversible. Confronted with all these poor people looking to their 'big sister' for miracles, the nun felt all the strength go out of her. Padmini gently mopped up the large beads of sweat that were coursing their way down her forehead and threatening to impede her vision. Rising above the nauseating smells and horrific sights, the young Indian girl supported her 'big sister' with her unfailing smile. The little girl's expression, it too born of suffering and poverty, revived the nun's courage whenever it faltered.

One day a woman deposited an extremely emaciated baby on the table. Sister Felicity had the idea of entrusting the tiny body with its dried up skin to Padmini.

'Take him and massage him gently,' she told her. 'That's all we can do.'

Padmini sat down on a jute sack in the alleyway and placed the child in her lap. She poured a little mustard oil on her hands and began to massage the small body. Her hands came and went along its upper torso and limbs. Like a succession of waves, they started on the baby's sides, worked across his chest and up to the opposite shoulder. Stomach, legs, heels, the soles of his feet, his hands, his head, the nape of his neck, his face, the wings of his nose, his back and his buttocks were successively stroked and vitalized, as if nourished by Padmini's supple, dancing fingers. The child suddenly began to gurgle for sheer bliss. 'I was dazzled by so much skill, beauty and intelligence,' Felicity would later say. 'In the depths of that slum I had just discovered an unsuspected power of love and hope. The people of Orya Bustee deserved the mercy of God.'

# 11

## 'A hand for the future'

Out of the thirty-eight countries on the planet where Union Carbide had hoisted its flag, no other had established such long-standing and warm links with the company as India. Perhaps this was due to the fact that for nearly a century the multinational had been providing a commodity as precious as air or water for hundreds of millions of Indians who had no electricity. Carbide's lamps brought light to the most remote villages on the Indian subcontinent. Thanks to the half a billion batteries made in its factories each year, the whole of India knew and blessed the American company's name.

The rich profits from this monopoly and the conviction that the country would one day become one of the world's great markets had induced Carbide to regroup all kinds of production under the aegis of an Indian subsidiary known as Union Carbide India Limited. So it was that the flag of this subsidiary company fluttered over fourteen factories. In India, Carbide manufactured chemical products, plastic goods, photographic plates, films, industrial electrodes, polyester resin, laminated glass and machine tools. The company also had its own fleet of seven trawlers on the Bengal coast, specializing in deep water shrimping. With an annual turn-over of 200 million dollars, Union Carbide India Limited was a successful example of the corporation's globalization policy. Of course, the New York multinational retained ownership of 51 per cent of the shares in its Indian subsidiary, which gave it absolute control over all its production and any new projects on Indian soil.

❧

In April 1962, the American management of Carbide revealed the nature and scope of these new projects with a full-page advertisement in *National Geographic Magazine*. Entitled 'Science helps to build a new India' the illustration was meant to be allegorical. It depicted a dark-skinned, em-aciated peasant working obviously infertile soil with the aid of a primitive plough drawn by two lean oxen. Two women in saris with a pitcher of water and a basket on their heads, surveyed the scene. In the background appeared the waters of a mighty river, the Ganges. Just beyond the sacred river, glittering with a thousand fires in the sunlight, there arose the gilded structures of a gigantic chemical complex with its towers, chimneys, pipework and tanks. Above it, in the upper half of the picture, a light-skinned hand emerged from the orange sky. Between thumb and index finger it was holding a test tube full of a red liquid, which it was pouring over the peasant and his plough. Carbide had no doubt drawn its inspiration from the scene on the ceiling of the Sistine Chapel in which Michelangelo portrays the hand of God touching Adam's to give him life. Under the heading, *A hand for the future*, the company delivered its message in the space of a single paragraph:

*Cattle working in the fields . . . the eternal River Ganges . . . elephants caparisoned with jewels . . . Today these symbols of ancient India coexist with a new vision, that of modern industry. India has built factories to strengthen its economy and provide its four hundred and fifty million people with the promise of a bright future. But India needs the technological knowledge of the Western world. That is why Union Carbide, working with Indian engineers and technicians, has made its scientific resources available to help construct a large plant to produce chemical products and plastic goods near Bombay. All over the free world, Union Carbide has undertaken to build plants to manufacture chemical products, plastic goods, gases and alloys. Union Carbide's collaborators are proud to be able to share their knowledge and skills with the citizens of this great country.*

This piece of purple prose concluded with an exhortation:

*Write to us for a brochure entitled:* The exciting world of Union Carbide. *In it you'll find out how our resources in the different domains of carbon, chemical products, gases, metals, plastic goods and energy continue daily to work new wonders in your life.*

'New wonders in your life!' This eloquent promise was soon to find a spectacular opportunity for fulfilment. It was at a time when India was trying desperately to banish the ancestral spectre of famine. After the severe food shortages at the beginning of the 1960s, the situation was at last improving. The source of this miracle lay with a batch of apparently unassuming grain imported from Mexico. Christened Sonora

63 by its creator, the American agronomist and future Nobel Peace prize-winner, Norman Borlang, the grain produced a new variety of high-yielding corn. With heavy ears that were not susceptible to wind, light variation or torrential monsoon rains, combined with very short stems that were less greedy, this fast growing seed made it possible to have several harvests a year on the same plot of land. It brought about a revolution, the famous Green Revolution.

This innovation suffered serious constraints, however. In order for the high-yielding seeds to produce the multiple harvests expected of them, they needed lots of water and fertilizer. In five years, between 1966 and 1971, the Green Revolution multiplied India's consumption of fertilizer threefold. But that was not all. The very narrow genetic base of high-yield varieties and the monoculture associated with them made the new crop ten times more vulnerable to disease and insects. Rice became the favourite target for at least a hundred different species of predatory insects. Most virulent were the small flies known as green leafhoppers. The stylets with which they sucked the sap from young shoots could destroy several acres of rice fields in a few days. In the Punjab and other states, the invasion of a form of striped aphid decimated the cotton plantations. Against this scourge, India had found itself virtually defenceless. In its desire to promote the country's industrialization, the government had encouraged the production of pesticides on a local basis. Faced with the enormity of demand, however, locally manufactured products had shown themselves to be cruelly inadequate. What was more, a fair number contained either DDT or HCH – hexa-chlorocyclohexane, substances considered so dangerous to flora, fauna and humans that a number of countries had banned their use.

Finding themselves unable to provide their peasants with a massive supply of effective pesticides, in 1966 Indian leaders decided to turn to foreign manufacturers. Several companies, among them Carbide, were already established in the country. The multinational was interested enough to dispatch a scout from their sales team to New Delhi. They sent the young Argentinian agronomical engineer, Eduardo Muñoz. After all, hadn't this engaging sales representative managed to convert the whole of South America to the benefits of Sevin? Muñoz promptly proved himself worthy of selection by inaugurating his mission with a masterstroke.

The legendary emperor Ashoka, who had spread the Buddha's message of non-violence throughout India, would have been amazed. On a winter evening in 1966, the hotel in New Delhi which bore his name welcomed the principal executives of Carbide's Indian subsidiary company together with about a hundred of the highest officials from the Ministry of Agriculture and the Planning Commission. These dignitaries had gathered for a banquet to celebrate the quasi-historic agreement signed that afternoon at the Ministry of Agriculture in front of a pack of journalists and photographers. The contract aimed to arm the Indian peasants against aphids and other insects destroying their crops. To this end, it provided for the immediate importation of 1200 tons of American Sevin. In return, Carbide undertook to build a factory to make this same pesticide in India within five years. Eduardo Muñoz had negotiated this agreement with a high-ranking official from the Ministry of Agriculture named Sardar Singh, who seemed in a great hurry to see the first deliveries arrive. He was, as his turban and the beard rolled round his cheeks indicated, a Sikh, originally from the Punjab. The peasants of his community had been the first victims of the marauding insects.

As chance would have it, the Carbide envoy was able to satisfy the hopes of his Indian partner sooner than anticipated. Discovering that a cargo of 1200 tons of Sevin, destined for farmers in the Nile valley, which had been devastated by invading locusts, was held up in the port of Alexandria because of the zeal of intrusive customs officers, the Argentinian managed to have the ship diverted to Bombay. A fortnight later, the precious Sevin was received there like a gift from heaven.

The euphoria subsided somewhat when it was discovered that the Sevin from the Egyptian ship was actually a concentrate that could not be used until it had undergone appropriate preparation. In their own jargon, specialists called this process 'formulation'. It consisted of mixing the concentrate with sand or gypsum powder. Like the sugar added to the active substance in a medicine to facilitate its consumption, the sand acts as a carrier for Sevin, making it possible to either spread or spray the insecticide according to the particular needs of the user. There was no shortage of small industrial units in India to carry out this transformation process. But Muñoz had a better idea. Carbide was going to make its Sevin usable itself, by building its own formulation factory. No matter that the *Industrial Development and Regulation Act* reserved the construction of this kind of plant for very small firms and only those of Indian nationality, he would easily get round the law by finding someone to act as a front-man.

In India, like anywhere else, there is no shortage of intermediaries, agents, *compradores*, prepared to act as go-betweens for any kind of business. One morning in June 1967, a jolly little man turned up in Eduardo Muñoz's office.

'My name is Santosh Dindayal,' he announced, 'and I am a devotee of the cult of Krishna.' Taken aback by this mode of introduction, the Argentinian offered his visitor a cigar.

'I own numerous businesses,' the Indian went on. 'I have a forestry development company, a scooter concession, a cinema, a petrol station. I've heard about your plan to build a pesticide factory.' At this point in his account, the man assumed a slightly mysterious air. 'Well, you see, it so happens that I have entrées all over Bhopal.'

'Bhopal?' repeated Muñoz, to whom the place meant nothing.

'Yes. It's the capital of the state of Madhya Pradesh,' continued the Indian. 'The state government is eager to develop its industry. It could well be useful for your project.'

Drawing vigorously on his cigar, the little man explained that the people running Madhya Pradesh had set aside an area for industrial development on vacant land north of the capital.

'What I'm proposing is that I apply in my name for a licence to construct a plant that can transform the Sevin your friends have imported into a product that can be used on crops. The cost of such an undertaking shouldn't be more than fifty thousand dollars. We can sign a partnership contract together. You do the work on the factory and then you can give me a proportion of the proceeds.'

The Argentinian was so pleased he nearly swallowed his cigar. The proposal was an excellent first step in the larger industrial venture he was counting on launching. It would provide an immediate opportunity to make Indian farmers appreciate what Sevin could do, and give the engineers in South Charleston time to come up with the large pesticide plant that the Indian government seemed to want to see built on its land. Suddenly, however, a question sprang to mind.

'By the way, Mr Dindayal, where is this town of yours, Bhopal?'

The Indian gave him an amused smile.

'In the very heart of India, dear Mr Muñoz,' he replied, pointing proudly to his chest.

✳

The heart of India! The expression excited the handsome Argentinian. Taking the Indian with him as navigator, he set off at once in his grey Mark VII Jaguar for the heart of the country. To him it was like arriving 'in a large village'. The industrial zone designated by the government lay just over a mile from the city centre and a little more than half a mile from the train station. In the past, the rulers of Bhopal had built the stables for their race horses there and the troops of the Sultaniya Infantry had used it as a parade ground. The dark colour of the soil accounted for the name of the place, Kali meaning black grounds. But the term may also have derived from the colour of the blood with which the earth was saturated. For it was here that, before thousands of spectators, the kingdom's executioners used to lop off the heads of those whom the Islamic *sharia*[1] had condemned to death, with a sabre.

The Argentinian was not likely to be put off by such morbid associations. Two days' exploring had convinced him. This town of Bhopal held all the winning cards: a central location, an excellent road and railway system, and abundant electricity and water supplies. As for the Kali Grounds, in his eyes they held yet another trump: the string of huts and hovels extending along their boundaries promised to provide a plentiful workforce.

---

[1] Islamic canonical law which is strictly applied.

73

# 12

## A promised land on the ruins of a legendary kingdom

'The large village' the Carbide envoy thought he had seen from inside his Jaguar was in fact one of India's most beautiful and arresting cities. But then Eduardo Muñoz had not had time to discover any of the treasures that had given Bhopal so high a position in India's cultural patrimony. Since 1722, when an Afghan general fell in love with the site and founded the capital of his realm there, Bhopal had been adorned with so many magnificent palaces, sublime mosques and splendid gardens that it was justifiably known as 'the Baghdad of India'. Above all, however, it was for its rich Muslim culture, its tradition of tolerance and the progressive nature of its institutions that the town held a distinguished place in India's history. The riches of Bhopal had been forged first by a Frenchman, and then by four innovatory female rulers, despite the *burkahs* that concealed them from the eyes of men. The commander-in-chief of the Nawab's armies, and subsequently the country's regent, Balthazar I de Bourbon, and after him, the Begums Sikander, Shah Jahan, Sultan Jahan and Kudsia had turned their realm and its capital into a model, admired as much in Imperial Britain as by many other African and Asian colonial countries. The four Begums had not only put an end to the isolation of their state by financing the advent of the railway line with their own funds, they had opened up roads and markets, built cotton mills, distributed vast acres of land to their landless subjects, set up a postal system without equal in Asia and introduced running water to the capital. Wanting to educate their people, they had introduced free primary instruction for everyone and promoted female emancipation by increasing the number of girls' schools. The last but one of these enlightened women rulers, Sultan Jahan Begum, had even created an institution which was revolutionary for the time, the Bhopal Ladies Club. There women were free to discuss their circumstances and their future. The same Begum had also given her female subjects the opportunity to go shopping with their faces uncovered by building the Paris Bazaar, a huge shopping centre reserved exclusively for women. There they could walk about with their faces uncovered because all the shopkeepers were women. Simply dressed and without bodyguards, the Begum herself liked to visit this emporium which was well-stocked with items imported from London and Paris. The British were unsparing in their respect for her. King George V invited her to his coronation and, in 1922, the Prince of Wales paid a visit for the inauguration of the Government Council for the Kingdom of Bhopal, a democratic institution unique in the princely India of that time. His visit was also intended to thank the Begum for having so generously emptied her privy purse and the state coffers, to support the British war effort. After all, she had sent her eldest son to represent Bhopal and fight alongside the Allied soldiers in the trenches of the Great War.

The magnificence of the kingdom and its prestigious capital expressed itself in many different ways. A great lover of literature and herself the author of several philosophical treatises, Shah Jahan Begum had attracted to her court the most distinguished scholars and learned men of India and of countries such as Afghanistan and Persia. The city had supplanted Hyderabad and Lahore as a beacon of renascent Islamic culture, with the Urdu language which is so rich in literature, as well as painting and music. Of all the expressions of this heritage, it was to poetry that the Begum contributed most. Reviving the tradition of the *mushaira*, evenings of poetry recital when the people could meet the greatest poets, she threw open the reception rooms of her palace to all and arranged for monumental performances on the Household Cavalry's Lal parade ground. Eyes bright with happiness, 60,000 to 80,000 poetry lovers, three-quarters of the town's population, used to come and sit on the ground right through the night to hear poets sing of suffering, joy and the eternal aspirations of the soul. 'Weep not, my beloved,' implored one of the Bhopalis' favourite refrains. 'Even if for now your life is but dust and lamentation, it already proclaims the magic of what lies ahead.'

Before she passed away, Kudsia Begum, last of the Queens of Bhopal, nevertheless expressed her regret that her subjects seemed more interested in poetry than industrial projects or affairs of state. Despite the efforts of the economic development agency she had created with the support of the British in the period between the world wars, very few firms came to Bhopal. Two spinning mills, two sugar refineries, a cardboard and a match factory – the sum total was a modest one. Nor did the ascendance to the throne of a male sovereign do anything to rectify matters. The Nawab Hamidullah Khan was a charming, cultivated prince but far more interested in the decoration of his palaces or the breeding of his horses than in the construction of blast furnaces or textile factories. While Mahatma Gandhi was going on hunger-strike to force the British out of the country, he was having a luxury tub installed on top of one of his hunting vehicles.

On 15 August 1947, the subcontinent's independence cast the maharajahs and nawabs of the Indian kingdoms into the oubliette of history. The upheaval was a stroke of good fortune for Bhopal, which found itself promoted to the capital of the vast province of Madhya Pradesh, taking in all India's central territories. Its selection, which spurred the city into an era of feverish development, was a consequence of precisely those trump cards which, twenty years later, impressed Carbide's representative. Buildings had had to be constructed to house the new province's ministries and administrative bodies, whole neighbourhoods had to be built in which to lodge the thousands of officials and their families: a university, several technical colleges, a hospital with 2000 beds, a faculty of medicine, shops, clubs, theatres, cinemas, restaurants. In the space of five years the population increased from 85,000 to nearly 400,000.

This rise had brought with it an influx of small and large firms from all over India. And now, as the chromed muzzle of a grey Jaguar had just intimated, America was about to step in where only yesterday the last Nawab and his guests had still been hunting tigers and elephants. So that, for the occupants of Orya Bustee, as for the hundreds of other immigrants who stepped off the trains each day looking for work, Bhopal, at the end of the 1960s, was the Promised Land.

77

# 13

## *A continent of 300 million peasants and 600 languages*

The city of the Begums greeted the government of Madhya Pradesh's decision as a gift from the gods. By assigning a five-acre plot of land on the Kali Grounds to the entrepreneur Santosh Dindayal, along with permission to build a factory to make pesticides, the government was offering the city all the opportunities that went with an industrial venture. Eduardo Muñoz was quick to pass on the glad tidings to his New York management, before rushing to the bar in Calcutta's luxury Grand Hotel, to sip champagne with his wife Rita and his colleagues. He then set about looking for a team to build the factory. By a stroke of incredible luck, he chanced upon the perfect trio: first Maluf Habibie, a frail Iranian chemical engineer with metal-rimmed spectacles, a specialist in formulation techniques for chemical products; then Ranjit Dutta, an engineer built like a rugby player, who had previously worked with Shell in Texas; finally, the only Bhopali, Arvind Shrivastava had just completed his degree in mechanical engineering. The three men worked furiously in the backroom of the petrol station belonging to Muñoz's Indian associate. In a fortnight they laid down the sketches for a plant. 'Plant' was a very grandiose name for a workshop to house the crushers, blenders and other equipment necessary for the commercial preparation of Sevin.

Like all of life's events in India, the first spade digging was marked with a ceremony. A pundit girdled with the triple thread of a brahmin came and chanted mantras over the hole dug out of the black earth. A coconut was brought for Arvind Shrivastava to slice with a bill-hook. The pundit poured the milk slowly onto the ground. Then the young man cut the flesh into small pieces, which he offered to the priest and the onlookers. The brahmin raised his hand and the workmen came forward and emptied their wheel-barrow full of concrete into the cavity. The gods had given their blessing. The venture could commence.

※

With no complicated pipework, no glistening tanks, no burning flares, no metal chimneys, the building that rose from the Kali Grounds bore no resemblance to the American monsters in the Kanawha Valley. In fact with its triple roof and line of small windows it looked more like a pagoda. Inside was a vast hangar with a range of conical silos mounted on grinding machines. This plant was to provide the Sevin concentrate imported from America with granular carrier agents adapted to the various methods of diffusion. The Sevin to be sprayed from the air over the huge plantations in the Punjab must be 'formulated' more finely than the packaged Sevin that was to be spread by hand by the small farmers of Madhya Pradesh or Bengal. Granular or fine-as-dust, either way the Bhopal Sevin promised to be a unique insecticide, less for its intrinsic qualities than for the carrier agent Muñoz's engineers had found for it.

To mystical India the River Narmada is the daughter of

the Sun. It is enough to behold it to achieve perfect purification. One night of fasting on its banks guarantees prosperity for hundreds of generations, and drowning in it wrests one from the cycle of reincarnations. By a fortuitous stroke of geography this sacred river flowed only twenty-five miles from Bhopal. Its banks were covered with a sand as mythical, according to the *Vedas*, as the waters they confined. Mixed with the pesticide from America, the sand from the Narmada would avenge the Nadar family and all the other peasants ruined by voracious insects. India was going to escape the ancestral curse of its famines.

❋

'It was the best Christmas present I'd ever had,' the turbaned Indian, who had bought the 1200 tons of American Sevin from Muñoz, would confide. The end of that year, 1968, saw the first delivery of insecticide, produced in the small factory in Bhopal, arrive in his ministry's warehouses: 131 tons to be sprayed over the cotton and cereal plantations of the Punjab. Once the requirements of his beloved Punjab had been satisfied, however, Sardar Singh was likely to find himself with seven or eight hundred tons of pesticide left on his hands. How could he ensure that other peasants in his country benefited from this surplus? He turned to Eduardo Muñoz for help.

'Your company sells more than 500 million batteries a year in this damn country,' he told him. 'Its agents range from the farthest reaches of the Himalayas and the backwaters of Kerala. Only an organization like yours can help me distribute my Sevin.'

The Argentinian raised his arms.

'My dear Mr Singh, a packet of insecticide is not as easy to sell as a pocket lamp,' he pointed out.

The Indian adopted a wheedling tone.

'My dear Mr Muñoz, what you personally have achieved in Mexico and Argentina, you will manage to achieve here too. I have every faith in you. Let's say no more about it: your smile tells me you will help me.'

The challenge was a colossal one. From behind the wheel of his Jaguar Muñoz had gauged the enormity and complexity of India. The country bore no resemblance to Mexico or even Argentina, which he had ended up knowing like the back of his hand. India was a continent with 300 million peasants speaking 500 or 600 different languages and dialects. Half of them were illiterate so they would be unable to read the label on a sack of fertilizer or a can of insecticide. Yet they were dealing with chemical products that were potentially fatal. Muñoz had been horrified by the number of accidents the newspapers reported in rural areas: lung damage, burns to the skin, poisoning. The victims were almost always poor agricultural labourers whose employers had not seen fit to provide them with protective clothing or masks. To improve the efficacy of their manure, many peasants mixed different products together and almost always with their bare hands. Some even tasted the mixture to make sure it had been made up properly. In the poorest villages where whole families lived in one room, the bag of insecticide frequently sat in one corner, insidiously poisoning them with toxic emissions. Women went to draw water, did the milking or cooked food with containers that had once held DDT. The result was an alarming increase in certain disorders. A journey through the Tamil Nadu region horrified the Union Carbide representative. In some areas known for their intensive use of phytosanitary products, the instances of cancers of the lung, stomach, skin, and brain defied counting. In the Lucknow region, half the labourers handling pesticides were suffering from serious psychological disorders as well as problems with their memory and eyesight. The worst part of it was the pointlessness of these sacrifices. Poorly informed peasants thought they could increase a product's effectiveness by doubling or tripling the manufacturer's recommended dosage. Their lack of understanding led many of them to ruin, sometimes even suicide. The newspaper headlines reported that the most popular method these desperate people used to kill themselves was swallowing a good dose of pesticide.

Despite his worries about the misuse of insecticides, Eduardo Muñoz responded to his Indian partner's appeal for help. He sent out the sales team for the batteries with the blue and white logo, to dispose of the surplus Sevin. Soon every single grocery, hardware shop and travelling salesman would be selling the American insecticide. This apparently generous gesture was not entirely devoid of self-interest. The Argentinian was counting on it to provide him with an accurate assessment of the Indian market's capacity to absorb pesticides. The information would be crucial when the time came to determine the size and production volume of the Indian plant that Union Carbide had promised to build.

❉

'Work with farmers, our partners in the field.' A tidal wave of notices bearing this slogan soon broke over the Bengali and Bihari countryside. They showed a Sikh in a red turban placing a protective hand on the shoulder of a poor old farmer with a face deeply furrowed with wrinkles. In his other hand, the knight in shining armour was brandishing a box of Sevin, the size of a packet of supermarket crackers. He was using it to point at an ear of corn. The copy read: *My name is Kuldip Chahal. I am an area pesticide technologist. My role*

*is to teach you how to make five rupees out of every rupee
you spend on Sevin.*

Eduardo Muñoz was all the more convinced: to convert
the Indian peasants to Sevin, he would need legions of Kuldip
Chahals.

# 14

## *Some very peculiar pimps*

The sudden invasion of concrete mixers, cranes and scaffolding, over the bleak horizon of the Kali Grounds, caused a major stir in the bustees. The blue and white logo flying in the vicinity of the mud huts was an even more magical emblem than the trident of the god Vishnu, creator of all things. To Eduardo Muñoz that flag hoisted high for all to see constituted a considerable victory. He had managed to persuade the New Delhi authorities that Union Carbide should no longer have to rely on an Indian intermediary to 'formulate' the Sevin concentrate. Rather it was to be able to operate openly, under its own name. In New Delhi, as elsewhere in the world, international big business invariably found its own ways and means.

As soon as the construction site opened, several *tharagars* laid siege to Belram Mukkadam's tea house. Carbide needed a workforce. Candidates came running and soon the drinks stall became a veritable job centre. Among the *tharagars*, Ratna Nadar recognized the man who had recruited him in Mudilapa to double the railway tracks. Ratna would have liked to have given him a piece of his mind, let him know just how bitter and angry he was, shout out that the poor were sick of others getting fat from the sweat of their labour. But this was not the moment. The dream of crossing the American multinational's mythical portals might just be about to come true.

'I pay twenty rupees a day,' announced the *tharagar*, exhaling the smoke from his *bidi*. 'And I supply a helmet and boiler suit, and one piece of soap a week.'

It was a small fortune for men used to feeding their families on less than four rupees a day. In gratitude they bowed to wipe the dust from their benefactor's sandals. Among them was the former leper, Ganga Ram. This would be the first job he had managed to land since leaving the wing for contagious diseases at Hamidia Hospital.

Next day at six o'clock, led by Mukkadam, all the candidates presented themselves at the gateway to the building site. The *tharagar* was there to check each worker's employment document. When it came to Ganga Ram's turn, he shook his head.

'Sorry, friend, but Carbide doesn't take lepers,' he declared, pointing to the two stumps of finger that were awkwardly gripping the sheet of paper.

Ganga Ram foraged in the waist of his *lunghi* for the certificate to show that he was cured.

'Look, look, it says there, I'm cured!' he implored, thrusting the paper under the *tharagar*'s nose.

The latter was inflexible. For Ganga Ram, the opportunity to don one of Carbide's boiler suits would have to remain a dream.

That evening, those who had been fortunate enough to receive the blue linen uniform took it home with them to present it to Jagannath, whose image watched over the neighbourhood from a small niche at the corner of the alleyway. Sheela, Padmini's mother, laid her husband's clothing at the deity's feet, placing a chapatti beside it and some marigold petals moistened with sugar water.

A few days later, Belram Mukkadam's chief informant brought a piece of news that restored the hopes of Ganga Ram and all the others who had not been hired.

'This building site is just the thin end of the wedge,' announced Rahul, the legless cripple. 'Soon, sahibs will be arriving from America to build other factories and they're going to pay wages higher than even Ganesh[1] could imagine.'

Rahul was one of the most popular characters in Orya Bustee. He got about at ground level on a plank with wheels, which he propelled with all the dexterity of a Formula 1 driver. With his fingers covered in rings, his long, dark hair carefully caught up in a bun, his glass bead necklaces, and his shirts with gaudy, geometric patterns, Rahul introduced a note of cheeky elegance to the place. He was always abreast of any news, the slightest whisper of gossip. He was the Kali Grounds' newspaper, radio and magazine. His smile and generous disposition had earned him the nickname 'Kali Parade Ka Swarga dut – the Angel of Kali Parade'.

That morning he was the bearer of another piece of news that was to appal all those gathered at the tea house.

'Padmini, Ratna and Sheela Nadar's daughter, has disappeared,' he announced. 'She hasn't been home for four days. She wasn't there this morning to help Sister Felicity with her clinic. Dilip, Dalima's son, says he and his friends lost her in the station at Benares.'

This piece of information sent everyone rushing to the Nadars' hut. In the bustee everyone shared their neighbour's misfortune.

❄

---

[1] The elephant-headed god, symbolizing prosperity.

That winter Dilip, Padmini and the gang of young ragpickers that worked the trains had been extending their expeditions further and further afield. They ventured beyond Nagpur, even as far as Gwalior, which prolonged their absence by two or three days. Hopping from train to train, they roved the dense railway network of northern India with ever more audacity. One of their most lucrative destinations was the holy city of Benares, situated some 375 miles away, to which trainloads of Hindus of all castes went on pilgrimage. They could be there and back in four days, which meant that if Padmini set out on a Monday, she could still be back in time for Sister Felicity's clinic, something she would not miss for the world. These long journeys were fraught with danger. One evening when she parted from her friends to run and buy some fritters, the train left without her. It was the last one that night. Alone in Benares' vast station overrun with travellers, vendors and beggars, Padmini panicked and burst into tears. A man wearing a white cap approached her and pressed a crumpled ten rupee note into the palm of her hands.

'Don't thank me, little one. I'm the one who needs you.' He invited the little girl to sit down beside him and told her that his wife had just been called away to Calcutta to look after her dying father.

'She won't be back for a few days and I'm looking for someone to take care of my three small children while she's away,' he explained. 'I live close by. I'll give you fifty rupees a week.'

Without giving her time to answer, the man scooped Padmini up by the armpits and carried her to a car parked in front of the station. Like all great pilgrimage centres, Benares played host to a fair number of dubious activities. One of the most flourishing was the prostitution of little girls. According to popular belief, deflowering a virgin restored a man's virility and protected him against venereal disease. The city's numerous pleasure houses relied on professional procurers. These procurers often bought girls from very poor families, notably in Nepal, or arranged fictitious marriages with pretend husbands. Sometimes they even simply abducted their victims.

Two men similarly dressed in white caps were waiting in the car for an adolescent girl to be 'delivered' to them. The vehicle took off at speed and drove for a long time before it stopped outside the grille of a temple. Twenty girls crouched inside the courtyard, guarded by more men in white caps. Padmini tried to escape from her abductors but she was forced into the courtyard.

In this city where every activity had sacred associations, some pimps tried to trick their young victims into believing they would be participating in religious festivals. Padmini's capture happened to coincide with the festival of Makara Sankrauti, celebrated on the winter solstice. Makara is the goddess of carnal love, pleasure and fertility.

The young captives were driven inside the temple where two pundits with shaven heads and chests encircled with the brahmin's triple cord, were waiting for them. 'That was the beginning of a nightmare that went on for two days and two nights,' Padmini recounted. Cajoling one minute, threatening the next, banging their gongs to punctuate their speech, performing all kinds of rituals at the feet of the numerous deities in the sanctuary, the men sought to break down their victims' resistance and prepare them for the work that awaited them. Fortunately Padmini did not understand the language they spoke.

Once their very peculiar training was over, the captives were taken under escort to Munshigang, Benares' brothel quarter, to be divided up between the various houses that had bought them. Padmini and two other little victims were pushed into one of the houses and taken to the first floor where a woman in her fifties was waiting for them.

'I'm your new mother,' declared the madam with a cajoling smile, 'and here are some presents that will turn you into proper princesses.'

She unfolded three different coloured skirts with matching blouses and showed them several boxes containing bracelets and necklaces, and make-up. The gifts were part of what the local pimps referred to as the breaking of the girls.

'And now, I'll go and get you your meal,' announced the procuress before leaving them and locking the door behind her.

It was now or never. Barely two yards separated the three little girls from the window of the room in which they were confined. Padmini made a sign to her companions, rushed over to the window, unbolted it, then jumped into the void. Her fall was miraculously broken by a fruit vendor's stall. She picked herself up, and seconds later was lost in the crowd. Her getaway had been so swift that no one had time to react. Following her instincts, the little girl ran straight ahead as fast as her legs would carry her. Soon she reached the banks of the Ganges and turned left along the *ghats*. In her flight she had lost her two companions but she was sure that they too had been able to escape. Jagannath had protected her. All she had to do now was find the station and climb aboard the first train for Bhopal.[2]

Two days later, as Dilip and his friends were making ready to slip aboard the Bombay Express, they suddenly caught sight of their little sister getting out of a carriage. They let

---

[2] According to the magazine *India Today*, 15 April 1989, over three thousand young girls a year are delivered into prostitution during the festival of Makara Sankrauti in the state of Karnataka alone.

out such shrieks of joy that the passengers rushed to the windows in curiosity.

'There you are,' said Padmini, taking a package from her bag. 'I've brought you some fritters.'

The boys bore her aloft in triumph, then took her home. News of her return, already broadcast by Rahul, brought hundreds of local residents rushing to her home.

# 15

## 'A plant as inoffensive as a chocolate factory'

An official letter from the Indian Ministry of Agriculture informed Eduardo Muñoz that the New Delhi government was granting Union Carbide a licence to manufacture 5000 tons of pesticide a year. This time it was not just a matter of adding sand to several hundred tons of concentrate imported from America, but permission to actually produce Sevin, as well as chemical ingredients, in India itself.

As usual, the Argentinian took his wife Rita and his colleagues to celebrate this latest success in the bar of the Grand Hotel in Calcutta. But as he raised his champagne glass to the success of the future Indian factory, he felt a nagging doubt. 'Five thousand tons, five thousand tons!' he repeated, shaking his head. 'I'm afraid our Indian friends may have been thinking a bit too big! A factory with the capacity for two thousand tons would be quite large enough for us to supply the whole of India with Sevin.'

The first sales figures for the Sevin 'formulated' in the small unit on the Kali Grounds were not very encouraging. This was the reason for Eduardo Muñoz's reluctance. Despite an extensive information and advertising campaign, the Indian peasants were not readily giving up familiar products like

HCH and DDT. The climatic variations of so immense a country with its late or inadequate monsoons and its frequent droughts that could suddenly reduce demand, meant that regular sales of the product could not be guaranteed. A salesman above all else, Muñoz had done his sums over and over again. His most optimistic predictions did not exceed annual sales of 2000 tons. Wisdom ordained that Carbide should limit its ambitions. Certain that he would be able to convince his superiors, he flew to New York. In his briefcase, meticulously classified by province, group of villages and sometimes even by village, were the results of his first sales effort. He hoped they would be enough to persuade his employers that they should modify their investment in India, even if it meant leaving room for eventual competitors. He was wrong. That journey to New York was to set the seal on the first act in a catastrophe.

❈

The Argentinian could never have imagined that his greatest adversary would be a man dead for twenty-one years. The whole of American industry continued to revere as a prophet the man who, shortly after the Second World War, had revolutionized relations between management and workforce. As an obscure employee in a Philadelphia bank, Edward N. Hay, who sported a short Charlie Chaplin-style moustache and oversleeves to protect his starched shirts, had seemed unlikely to leave much of a legacy except memories. His obsessive ideas, however, would make him as famous a figure in the industrial world as Frederick Taylor, the man who developed the theory of scientific management of factory work. According to Edward N. Hay, the workforce did not receive the attention they warranted. Starting from this premise, he had devised a point system to evaluate every job done in a company. The idea was immediately adopted by a number of branches of American industry. By the end of the 1960s Union Carbide was one of the most enthusiastic users of his methods. All of its industrial projects were automatically given a point value, according to a system which determined the importance, size and sophistication of any installations to be constructed. The more numerous and complex the project, the higher the number of points. Because each point corresponded to a salary advantage, it was in the interests of the engineers assigned to planning and implementing any industrial project to see that, right from the outset, it was given the maximum number of points possible.

***

'I realized at once, I didn't stand a chance,' Eduardo Muñoz remembered. 'Even before they heard what I had to say, the management committee made up of all the divisional heads and the principal members of the board of directors had rallied enthusiastically in support of the Indian proposal.'

'India has a market of three hundred million peasants,' the chairman immediately declared.

'Five hundred million soon,' one of the directors added.

'Don't you worry, Eduardo, we'll sell our five thousand tons, and more!' the chairman insisted with the unanimous approval of all those present. He continued: 'To show you just how much faith we have in this project, we're allocating it a budget of twenty million dollars.'

'An extravagant sum that Mr Hay's point system was going to spread in a manner advantageous to everyone,' Muñoz said after meeting the South Charleston engineers in charge of laying the plans for the factory. These men were high-level chemists and mechanics, respected leaders in the field of manufacturing processes, in charge of reputable projects, in short they were the élite of the workforce at Union Carbide's technical research centre in South Charleston. 'But they were all little dictators,' Muñoz went on to say. 'They were obsessed with just one idea, that of using their twenty million dollar bounty to create the most beautiful pesticide plant India would ever be able to pride itself on.'

Showing them his documents, the Argentinian tried desperately to explain to the South Charleston team the distinctive characteristics of the Indian market. His line of reasoning left them cold.

'The Indian government's licence is for an annual production of five thousand tons of pesticide. So we have a duty to build a plant to produce five thousand tons,' the project's chief engineer interjected in a cutting voice.

'Clearly my commercial arguments were of no concern to those young dogs,' Muñoz would later remember. 'They weren't bound by any obligation to make a profit. They were simply itching to plant their flares, reactors and miles of piping in the Indian countryside.'

In the face of such obstinacy, the Argentinian sought a compromise.

'Wouldn't it be possible to proceed in stages?' he suggested. 'That is to say, to start by building a two thousand ton unit, which could then be enlarged if the market proved favourable?'

The question brought sarcasm from his audience.

'My dear Muñoz,' the project chief went on, 'you must appreciate that engineering work for this type of factory requires that we establish the size of production envisaged from the outset. The reactors, tanks and controlling mechanisms of a plant that manufactures two thousand tons of Sevin are not of the same calibre as those of a factory two and a half times larger. Once a production target has been set, it can't be changed.'

94

'I take your point,' conceded Muñoz, trying to be tactful. 'Especially as I imagine it's possible to slow down production in a factory that is larger than necessary to adapt production to demand?'

'That's exactly right,' the project chief agreed, pleased to see the discussion ending with consensus.

Alas, this consensus was only an illusion.

The Argentinian still had plenty of issues to take up with the men from South Charleston. The most important one had to do with the actual conception of the Indian factory. The Institute factory near Charleston which had been designed to produce 30,000 tons of Sevin a year and which was to serve more or less as a model, functioned continuously, the production line ran day and night. In order to maintain this continuity, considerable quantities of MIC, methyl isocyanate, had to be manufactured and stored. At the Institute plant, three tanks made out of high resistance steel and fitted with a complex refrigeration system, stored up to 120 tons of MIC.

To Muñoz's way of thinking, stocking such a quantity of this highly dangerous product might be justifiable for a factory like the one at Institute which ran twenty-four hours a day, but not in a much more modest plant where production was carried out as the need arose. For his own peace of mind the Argentinian went to Bayer in Germany and to the French Littorale factory near Béziers. Both companies handled MIC. 'All the experts I met went through the roof when I told them our engineers intended to store twenty-two or twenty-six thousand gallons of MIC in the tanks of the proposed Bhopal plant,' Muñoz recounted. 'One German told me: "We only produce our methyl isocyanate as needed. We'd never risk keeping a single litre for more than ten minutes." Another added: "Your engineers are out of their minds. They're putting an atomic bomb in the middle of

95

your factory that could explode at any time." As for the Béziers engineers, the French government had quite simply forbidden them to stock the MIC they used other than in a small number of twenty gallon drums that they imported directly from the United States as required.'

Shaken by the unanimity of these statements, the Argentinian returned to South Charleston to try and convince Carbide that it should modify its plans for the proposed Bhopal plant. As a preferable alternative to continuous production involving the permanent storage of tens of thousands of gallons of potentially fatal materials, he suggested the 'batch' production of MIC to meet production line requirements as they arose, such as was used at Béziers. This system eliminated the need to keep large quantities of dangerous substances on site. 'I quickly realized that my proposal ran counter to American industrial culture,' Muñoz recalled. 'They love to produce things in a continuous flow and in huge quantities. They're besotted with enormous pipes running into giant tanks. That's how the whole of the petrol industry and many others operate.'

Nonetheless, the South Charleston team wanted to allay the visitor's fears.

'The numerous safety systems with which this type of plant is equipped enable us to control any of the MIC's potentially dangerous reactions,' the project leader assured him. 'You have absolutely no need to worry, dear Eduardo Muñoz. Your Bhopal plant will be as inoffensive as a chocolate factory.'

❀

Other problems awaited the Argentinian on his return to India. His next priority was to find a site for the prospective factory. His superiors in New York and Charleston had agreed upon the choice of Bhopal, which was already home to the Sevin 'formulation' unit. But the new site would have to be of a completely different size. The new plant would be a hydra-headed monster. There would be the unit producing alpha naphtol, one for carbon oxide, one for phosgene, one for methyl isocyanate etc. Alongside these installations with their control rooms, works and hangars, the 'beautiful plant' would also have a collection of administrative buildings, a canteen, an infirmary, a decontamination centre and a fire station as well as a whole string of surveillance posts. All together it would need over a hundred and twenty acres all in one block and an infrastructure capable of supplying the enormous quantities of water and electricity necessary.

The Kali Grounds met all these conditions. But the Argentinian was against the site. 'I'd lost the battle over my conception for the factory,' he would say later. 'But at least I could try and stop it being built too close to areas where people were living.' The leaders of the Madhya Pradesh government rolled out the red carpet. The arrival of a multi-national as prestigious as Union Carbide was an extraordinary godsend for the town and the region. It meant millions of dollars for the local economy and thousands of jobs. Ratna Nadar, along with all the other residents of the bustees, would be kept in work for years.

Together with Muñoz, the Carbide team from New York examined several sites suggested by the authorities. None of them were really satisfactory. In one place the water supply was inadequate; in another it was electricity that was wanting; elsewhere the ground was not firm enough to take the weight of construction. That was when the residents of Orya Bustee and its neighbouring bustees witnessed cars mysteriously coming and going from the Kali Grounds. The vehicles frequently paused to let their occupants out. This activity went on for several days, then stopped. The envoys from New York had finally overcome Muñoz's reservations. Of course the Kali Grounds, next to the formulation works, was the right place to build the plant. As for any risk to those living nearby if an accident were to occur, the New York envoys reassured Muñoz that his fears were totally unfounded.

'Eduardo, if this plant is built as it should be, there will be no danger,' declared the man in charge.

'Take New York, for example,' interjected his assistant. 'Three airports surrounded by skyscrapers: la Guardia, JFK and Newark. Planes take off every minute and logically they should crash into the buildings whenever it's the least bit foggy, or collide with one another.'

'And yet,' his boss went on, 'New York's airports are the safest in the world. It will be the same in Bhopal.'

Muñoz had little choice but to agree. He and his colleagues presented themselves at the Madhya Pradesh government offices and submitted in due form their request for a hundred and twenty acre plot of land on the Kali Grounds. The piece of land in question was to adjoin the five acres of the 'formulation' works. According to municipal planning regulations, no industry likely to give off toxic emissions could be set up on a site where the prevailing wind might carry effluents onto densely populated areas. This was the case with the Kali Grounds where the wind usually blew from north to south, in other words onto the bustees, then the railway station and finally, onto the overpopulated parts of the old town. The application should have been turned down. But the Union Carbide envoys had taken good care not to mention that their proposed factory would be making its pesticides out of the most toxic gases in the whole of the chemical industry.

Clearly, Indira Gandhi had no great affection for her country's maharajahs and nawabs. When the British left, her father, Pundit Nehru, and the leaders of the Indian Independence movement, had taken their power away from them. She had then proceeded to confiscate their last remaining privileges and possessions. Eduardo Muñoz saw their persecution as a providential gift. The imaginative Argentinian dreamt of building in Bhopal, in tandem with the pesticide plant, a research centre along the lines of the American Boyce Thompson Institute where Sevin had been invented. After all, the Indian climate, the diseases and insects that damaged its crops were all factors associated with a particular environment. A research centre working on the ground in India might come up with a new generation of pesticides better suited to the country. It would be an opportunity for the proposed plant to diversify production and, who knows, perhaps one day hit the jackpot with new molecules that could be exported all over Asia. The idea went down well. Indian researchers and technicians would work for salaries ten or twelve times less than those of their American colleagues. All that was missing was a location. When Muñoz discovered that the brother of the last Nawab, threatened with government expropriation, was seeking to sell his Jehan Numa palace, he leapt at the chance. Rising magnificently from Shamla Hill, one of the seven hills surrounding the city, the edifice dominated the town. Its park, made up of several acres of tropical vegetation, rare trees, shrubs and exotic blooms, formed a sumptuous oasis of coolness, colour and scent. The building would probably have to be demolished, but the estate was vast enough to accommodate research laboratories, planetaria, greenhouses and even a luxurious guest house for passing visitors. In the conviction that an Indian would handle the transaction more adeptly than he, the Argentinian placed his assistant, Ranjit

99

Dutta, in charge of negotiations. They were hustled through. Three days later this jewel of Bhopal's ancestral patrimony fell into the clutches of the American multinational for the derisory price of 1,100,000 rupees, approximately $65,000.[1]

---

[1] When the palace was demolished, the magnificent Venetian crystal chandeliers that illuminated the feasts held by the Nawab were taken down and stored in packing-cases. The authors have never been able to recover any trace of them.

# 16

## A new star in the Indian sky

The India of *sadhus*[1] clad only in the sky, of sacred elephants caparisoned in gold; the India of devotees of God praying in the waters of the Ganges; the India of women dressed in multi-coloured saris planting rice in the south or picking leaves in the tea plantations of the Himalayas; the immemorial India of the worshippers of Shiva, Muhammad and Buddha; the India which had given the world prophets and saints such as Gandhi, Rabindrannath Tagore, Ramakrishna, Sri Aurobindo and Mother Teresa; the India of our fantasies, myths and dreams, had yet another face. The country that by the end of the century would number one billion inhabitants was already, by the 1960s, a fully developing industrial and technological power.

No one was more surprised by this than the small group of American engineers sent to Bombay by Union Carbide in 1960 to build a petrochemical complex. The venture united two vastly different cultures, with the magic of chemistry as their only common denominator. So constructive did this encounter prove to be that Carbide took on a whole team

---

[1] Sadhu: Hindu ascetic.

of young Indian engineers to inject new blood into the veins of the mighty American company. All those young men thought, worked and dreamed in English. They came from great schools like the Victoria Jubilee Technical Institute of Bombay founded by the British, or those created by the young Indian republic, such as the Madras Technical College, the Indian Institute of Science in Bangalore and the prestigious Rajputi College in Pilani. Some were graduates of eminent Western universities, such as Cambridge, Columbia University in New York or the Massachusetts Institute of Technology in Boston. Hindu, Muslim, Sikh or Christian, whatever their religion, they shared the same faith in science. The mantras they chanted were the formulae for chemical processes and reactions. Living in an economy that modelled itself on pro-tectionism and socialism, they were only too delighted to have prized open the door of a Western company where they could show off their talents, know-how, imagination and creativity. It was Carbide's genius to play this Indian card and involve the cream of local talent in its designs for indus-trial globalization.

'One good thing about this recognition was that it dis-pelled the archaic image many Westerners had of our country,' the engineer Kamal Pareek would say. Son of an Uttar Pradesh lawyer, a former student of the celebrated Pilani college, tennis champion and American film buff, at twenty-three this baby-faced young man was the embodi-ment of the youthful Indian energy Carbide was keen to harness. 'We Indians have always been particularly sensitive to the potential of the transformation of matter,' he confided. 'Our most ancient Sanskrit texts show that this sensitivity is part of our culture. We have a long-standing tradition of producing the most elaborate perfumes. Since the dawn of time our Ayurvedic medicine has used chemical formulae borrowed from our plants and minerals. The mastery of chemical elements is part of our heritage.' Pareek liked to illustrate what he had to say with examples. 'In Rajasthan there is a tribe of very backward people called the Bagrus,' he recounted. 'They make dyes for fabrics out of indigo powder, which they mix with crushed horn from horses' hoofs. To that they add pieces of bark from a tree called the ashoka and the residues of ant-infested corn. These people who have had no education, who are completely ignorant of the chemical phenomena operating at the heart of their concoctions, are on a par with the foremost chemists. Their dyes are the best in the world.'

❋

The first chemical plant built by Carbide in India was inaugurated on 14 December 1966. The blue and white flag hoisted into the sky over the island of Trombay was symbolic. A few miles from the spot where, four and a half centuries earlier, the galleon *Hector* had unloaded the first British, it embodied the desire of a new set of adventurers to make India one of the platforms for its industrial worldwide expansion. After the island of Trombay, it was Bhopal's Kali Grounds that were to see the same flag fly outside a highly sophisticated plant. The potentially deadly toxicity of its intended products had, however, sown doubt in the minds of some of the New York management. Was it wise to hand over technology as complex and dangerous as that associated with methyl isocyanate to a Third World country? In the end the excellent qualifications of the Indian engineers recruited for the Trombay factory allayed their fears. The Indians were invited to South Charleston to have some input into the plans for the Bhopal plant, an experience which the young technician, Umesh Nanda, son of a small industrialist in the Punjab, would never forget.

103

'Encountering the Institute Sevin plant was like being suddenly projected into the next millennium,' he recalled. 'The technical centre designing the project was a hive, inhabited by an army of experts. There were specialists in heat exchangers, centrifugal pumps, safety valves, control instruments and all the other vital parts. You had only to supply them with the particulars of such and such an operation to receive in return descriptions of and detailed plans for all the apparatus and equipment necessary. To mitigate the dangerous nature of the substances we were going to be using in Bhopal, bulky Safety Reports told us about all the safety devices installed at Institute. For weeks on end, we made a concerted effort with our American colleagues to imagine every possible incident and its consequences: a burst pipe, a pump breaking down, an anomaly in the running of a reactor or a distillation column.'

'It was a real pleasure working with those American engineers,' Kamal Pareek said for his part. 'They were so professional, so attentive to detail, whereas we Indians often have a tendency to overlook it. If they weren't satisfied, they wouldn't let us move on to the next stage.'

The pursuit of perfection was Carbide's hallmark. The company even brought over a team of Indian welders to familiarize them with the special alloys resistant to the acids and high temperatures with which they'd be working. 'Going to America to learn how to make up alloys as temperamental as Inconel, Monel or Hastelloy, was as epic a journey as flying off in Arjuna's chariot to create the stars in the sky,' marvelled Kamal Pareek.

The stars! Eduardo Muñoz, the magician behind the whole venture, could give thanks to the gods. The pesticide plant he was going to build on the Kali Grounds might not be exactly what he had dreamed of, but it did promise to be a new star in the Indian sky.

At the beginning of the summer of 1972 Carbide dispatched all the plans for the factory's construction and development to India. This mountain of paperwork was the finest gift American technology could offer a young industry in the developing world. Unfortunately, the actual plant was an imperfect replica of the Charleston installation. For reasons of economy Bhopal's 'beautiful plant' would not be provided with all the safety equipment and security systems the engineers in South Charleston had envisaged. The precise reasons for these economies would remain obscure. It seems that the sales of the Sevin 'formulated' in Bhopal did not reach the hoped-for level. Disastrous climatic conditions and the appearance on the market of a competing and less costly pesticide may have accounted for the reduction in sales. Because Indian law severely restricted the involvement of foreign companies in their local subsidiaries, Union Carbide India Limited suddenly found itself having to cut the factory's budget. Experts claimed, however, that none of these cutbacks would diminish the overall safety of the plant.

※

Four years late, the giant puzzle designed in South Charleston was created piece by piece in Bombay, then transported to Bhopal for assembly.

'Taking part in the project was like embarking on a crusade,' John Luke Couvaras, a young American engineer, said enthusiastically. 'You had to put yourself into it, body and soul. You lived with it every minute of the day and night, even when you were a long way from the works. If, for example, you were installing a distillation tower you'd fussed over lovingly, you were as proud as Michelangelo might have been of the ceiling in the Sistine Chapel. You kept an eye on it to make sure it went like clockwork. That kind of venture forced you to be vigilant at all times. It exhausted you, emptied you. At the same time you felt happy, triumphant.'

# 17

# 'They'll never dare send in the bulldozers'

American or Indian, none of the engineers and technicians working on the Kali Grounds could ever have
envisaged all the suffering, trickery, swindling, love, faith
and hope that was life for the mass of humanity occupying
the hundreds of shacks around the factory. In any poverty-
stricken area there were times when the worst existed alongside the best, but the presence of figures like Belram
Mukkadam managed to transform these patches of hell into
models for humankind. He was a devout Hindu, but there
were many Muslims, Sikhs and animists who made a stand
with him, and perhaps most remarkably, an Irani. The Iranis
with their light skins and delicate features formed a small
community of some 500 people. Their forefathers had come
to Bhopal in the 1920s, after an earthquake destroyed their
villages in Baluchistan, on the borders of Iran. Now, their
leader was an august old man with honey-coloured eyes, by
the name of Omar Pasha, invariably dressed in a *kurta*[1] and
cotton trousers. He lived with his sons, his two wives and
his lieutenants in a modern three-storeyed building on the

_____

[1] Long tunic worn over trousers.

edge of Orya Bustee. Three times a week, he would tear himself away from his comfortable life to take the sick from the three bustees to Hamidia Hospital. Driving those poor wretches through traffic that terrified them, then steering them along hospital corridors into packed waiting rooms was no mean feat. But without an escort the poor had little chance of being examined by a doctor. And even if they were lucky enough, they would not have been able to explain what was wrong with them or understand the recommended treatment. The majority of them spoke neither Hindi nor Urdu but one of the innumerable regional dialects or languages. Omar Pasha demanded that the slum dwellers be treated like human beings and made sure they actually received the medicines they were prescribed. Yet this Saint Bernard was one of Bhopal's most notorious godfathers. It was he who controlled the traffic in opium and *ganja*, the local hashish, as well as the brothels in the Lakshmi Talkies district; he ran the gambling, especially *satha*, which consisted of betting on the daily share-price of cotton, gold and silver.

He was also head of a real estate racket that made him one of the richest property owners in the town. To assure himself of the political support necessary to indulge in such activities he gave generously to the Congress Party, for which he was one of the district's most active electoral agents. The voting slips of Orya Bustee, Chola and Jai Prakash were in his hands. Good old Omar Pasha! His enormous fingers and powerful biceps testified to the fact that he had been a boxer and wrestler in his youth. With advancing age he had turned to another sport: cock-fighting. He bought his champions in Madras and fed them himself on a mixture of egg yolk, clarified butter, and crushed pistachio and cashew nuts. Before every fight he would rub them down 'like a boxer before a match', as he would say with a hint of nostalgia.

His ten cocks roamed freely about the floors of his house, watched over by bodyguards for, depending on its prize-list, each one was worth between twenty and thirty thousand rupees – a sum Padmini's father could not hope to earn in ten years of hard labour.

✳

The area was home to a host of other colourful people. The dairyman Karim Bablubhai distributed a portion of the milk from his seventeen buffalo cows to children with rickets. He dreamed of Boda, the young orphan girl from Bihar whom he had just married, giving him an heir. The yellow-robed sorcerer Nilamber, who exorcized evil spirits by sprinkling his clients with 'country liquor', had promised him that this dream would come true provided Boda performed a *puja* at the sacred *tulsi* every day. There was also the Muslim shoe-maker Mohammed Iqbal, whose hut on alleyway no. 2 exuded an unbearable smell of glue, and his associate Ahmed Bassi, a young tailor of twenty, who was famous for embroidering the marriage saris for the rich brides of Bhopal with gold thread. The Carbide engineers would have been very surprised to discover that in the sheds made out of planks, sheet metal and bamboo that they could see from the platforms of their giant Meccano piece, men in rags were producing masterpieces. The shoemaker and the tailor, like their friend Salar the bicycle repairman in alley no. 4, were always ready to respond to Belram Mukkadam's call. In the bustee no one ever declined to give him a helping hand.

Then there was Hussein, the worthy mullah with the small grey goatee beard, who taught the children *suras* from the Koran under the porch roof of his small, mud-walled mosque in Chola. And the old midwife Prema Bai, crippled by a childhood bout of polio, who dragged herself from hut to hut in her white widow's clothing. Yet her luminous smile outshone her suffering. In one corner of her hut, under the little altar where an oil lamp burned, day and night, before the statuette of Ganesh, god of good fortune and prosperity, the old woman carefully laid out the instruments that made her an angel of the bustee: a few shreds of sari, a bowl, two buckets of water and the Arabian knife she used to cut the babies' umbilical cords.

Who would have believed it? America and all its advanced technology was moving into the middle of a ring of hovels, without knowing anything at all about those who washed up against the walls of its installations like the waves of an ocean. No expatriate from Charleston, or Indian engineer moulded to Carbide's values, would ever be curious to know about the universe inhabited by those thousands of men, women and children living but a stone's throw away from the three methyl isocyanate tanks they were in the process of assembling.

One day, however, Carbide did pay a visit to the *terra incognita* that bordered on the Kali Grounds. 'People thought the end of the world had come,' Padmini's father used to say. The occupants of the bustees heard a plane roar overhead. The aircraft described several circles, skimming so low that they thought it would decapitate the Chola mosque's small minaret. Then, in a flash, it disappeared into the setting sun. This unusual apparition provided food for furious discussion at the tea house. Rahul, who always liked to think of himself as well-informed, claimed that it was 'a Pakistani plane come to pay homage to the fine factory that the Muslim workmen were building in their town of Bhopal.'

✤

The plane that appeared over the Kali Grounds was indeed the bearer of a homage, but not the one the legless cripple had imagined. The twin-engine jet plane *Gulf Stream II* that put down, on 19 January 1976, at Bhopal's airport, displayed on its fuselage a company crest with gilded wings marked with the initials UCC. It was transporting the Union Carbide Corporation's Chairman, a tall strapping fellow of fifty with white hair and a youthful air. A graduate of the prestigious Harvard Business School, and a former US navy reserve officer, Bill Sneath had climbed every rung of the multinational before becoming its chief in 1971. He was accompanied by his wife, an elegant young woman in a Chanel suit, and an entourage of high officials in the company. They had all come from New York to inaugurate the first phytosanitary research and development centre built by Carbide in the Third World.

The architecture of this ultra-modern edifice, with its façades dripping with glass, was inspired by the American research centre in Tarrytown. Built on the site of the palace that had belonged to the last Nawab's family and had been bought by Eduardo Muñoz, it very nearly never came into being. While digging the foundations, the masons had uncovered the skeleton of a bird and several human skulls. Word had then gone round that they belonged to three workmen who had mysteriously disappeared during the construction of the palace in 1906. This appalling omen immediately put the masons to flight. To entice them back, Eduardo Muñoz had had to resort to strong measures. He had tripled their salaries and arranged for a *puja* to lift the evil spell. When Bill Sneath arrived, the centre already comprised several laboratories, in which some thirty researchers were working, and greenhouses, in which many varieties of local plants were being grown.

The Indian minister of science and technology, the highest authorities of the state of Madhya Pradesh and the city, and all the local dignitaries from the chief administrator to the most senior police officer, gathered round the Sneaths, the Muñozes and the board of directors of Carbide's Indian subsidiary for the ceremony that sealed, in grandiose fashion, the marriage between the New York multinational and the City of the Begums. Before his speech, Bill Sneath received from one of the sari-clad hostesses the *tilak* of welcome, a dot of red powder on the forehead symbolizing the third eye which can see beyond material reality. The eyes of Carbide's Chairman surveyed with pride the vast concrete and glass block of the magnificent research centre. A few moments earlier they had visited the construction site of the future Sevin plant, where towers, chimneys, tanks and scaffolding were beginning to emerge from the earth of the Kali Grounds. Wearing helmets bearing their names, Bill Sneath and his wife had toured the building site pursued by photographers. In his hand, Sneath triumphantly brandished a packet of Sevin 'formulated' on site.

What the American would not see that winter was the jumble of huts, sheds and hovels that fringed the parade ground and grew like the swelling of a malignant cancer. Most of the men who lived there with their families made up the workforce for Carbide's various building sites. They had been invited to the inauguration of the research centre. The memento each had been handed by his supervisor was not, perhaps, very valuable, but for Padmini's father and all those living in homes with no lighting, a torch and three batteries stamped with the blue and white logo was a most precious gift.

❋

The gift that Sanjay Gandhi, the younger son of India's Prime Minister, had in store for several million of his country's

112

poor that same winter, was very different. Taking advantage of the state of emergency imposed by his mother to establish her power and muzzle the opposition, the impetuous young man had taken it into his head to clean up India's principal cities by ridding their pavements and suburbs of 'encroachments', in other words squatters. It was alleged that in towns approximately one tenth of occupiable land was taken up by people with no title deeds. This was the case with the bustees on the Kali Grounds. The sanitary conditions there were so abominable and the risk of epidemic so flagrant that the municipal authorities had often considered destroying the neighbourhood. But the local politicians, more concerned about keeping votes in the next election than getting rid of islands of poverty, had always opposed such radical action. Strengthened by the support of the beloved son of the all-powerful Indira, this time, however, Bhopal's municipal leaders had decided to take action.

One fine morning, two bulldozers and several truckloads of policemen burst onto the esplanade in front of the tea house. The officer in charge of the operation clambered onto the leading truck, armed with a loudhailer.

'People of Orya Bustee, Jai Prakash and Chola! By order of Sanjay Gandhi, central government and the city authorities, I am charged to warn you that you must leave the sites you are occupying illegally,' he declared. 'You have one hour in which to vacate the place. After that deadline, your huts will be destroyed and all those remaining will be apprehended and taken by force to a holding camp.'

'Oddly enough, the appeal didn't provoke any reaction at first,' Ganga Ram, the former leper, recalled. People formed a silent mob in the alleyways. It was as if the threats shouted through the loudspeaker had stunned them. Then suddenly, one woman let out a howl. With that all the other women began to shriek as if their entrails were being torn out. The sound was terrifying. Children came running from all sides like crazed sparrows. The men had rushed to the tea house. Rolling along on his plank on wheels, Rahul, the legless cripple, rounded everyone up. Old women went to take offerings and incense sticks to the statues of the gods in the district's various shrines. In the distance the inhabitants of the bustee could hear the bulldozers roaring like wild elephants eager to charge. That was when Belram Mukaddam appeared armed with his stick. When he began to speak outside the tea house, he seemed very sure of himself.

'This time the shits have come with bulldozers,' he thundered. 'Even if we lie down in front of their caterpillars, they won't stop at crushing us to pulp.'

He paused after these words, as if thinking. He fiddled with his moustache. 'You could see things were churning away in his head,' Ganga Ram said. He started to speak again.

'We do have one way of blocking those scum,' he went on, swiping at the air several times with his cane. He seemed to be savouring what he was about to say. 'My friends, we're going to change the names of our three bustees. We're going to call them after the much-loved son of our high priestess, Indira. We're going to call them the "Sanjay Gandhi bustees". They'll never dare, yes, I can assure you, that they'll never dare send in their bulldozers against a neighbourhood named after Sanjay!'

The keeper of the tea house then pointed his stick at a rickshaw waiting outside the entrance to the Carbide worksite.

'Ganga!' he directed the former leper, 'jump in that rattletrap and hurry to Spices Square! Get them to paint a big banner marked WELCOME TO THE SANJAY BUSTEES. If you get back in time, we're saved!'

114

Just as the apostle of the Kali Grounds had so magnificently predicted, the banner strung between two bamboo poles at the mouth of the road leading to Orya Bustee caused the tide of policemen and the bulldozers to stop dead in their tracks. That piece of material with the first name of Indira Gandhi's son written on it in imposing red letters was more powerful than any threat. The residents could go back to their huts without fear. Destiny would take it upon herself to crush them in a different way.

# 18

## *Wages of fear on the roads of Maharashtra*

The deadly cargo had arrived. As soon as he received the telex, the Hindu engineer Kamal Pareek alerted his assistant, the Muslim supervisor Shekil Qureshi, a chubby, thickset fellow of thirty-six. They packed the protective suits, gloves, boots, masks and helmets provided for special operations, into two suitcases, and caught the plane to Bombay. Their mission: to escort two trucks loaded with sixteen drums, each containing forty-four gallons of MIC over a distance of 530 miles. The Bhopal factory was not yet ready to make the methyl isocyanate required to produce Sevin. So its management had decided to have several hundred barrels brought over from Institute in the United States.

'Ships transporting toxic substances had to report to Aji Bunder,' Kamal Pareek recounted. 'It was a completely isolated dock at the far extremity of the port of Bombay. People called it "the pier of fear".'

Pareek watched with a certain amount of apprehension as the palette of drums dangled in mid air on the end of a rope. The crane was preparing to deposit its load in the bottom of a barge moored alongside the ship, which was then to transport the drums to the pier. Suddenly the engineer froze. Bubbles of gas were escaping from the lid of one of the containers. The ship's commander, who had spotted the leak, shouted to the crane operator:

'Tip your palette into the drink! Quickly!'

'No! Whatever you do, don't do that!' intervened Pareek, gesticulating frantically for them to stop the manoeuvre. 'One drum of MIC in the water, and the whole lot will go up!' Turning to the skipper of the barge, he shouted: 'Run for it, otherwise you and your family have had it!'

The skipper, a little bare-chested man, surrounded by half a dozen kids, shook his head.

'*Sahib*, my grandparents and my parents lived and died on this barge,' he replied. 'I'm ready to do the same.'

Pareek and Qureshi swiftly pulled on their protective suits, masks and helmets. Then, armed with several fat syringes full of a special glue, they jumped onto the bridge of the boat where, with infinite caution, the crane had deposited the palette. Clusters of yellow bubbles were still oozing from the damaged cover of one of the containers. The two men carefully injected the glue into the crack and managed to stem the leak. 'I heaved the biggest sigh of relief of my life,' Pareek would recall.

One hour later, the sixteen drums marked with the skull and crossbones were loaded aboard the two trucks. An agonizing journey was about to begin. Caught up in the chaos of *tongas*[1], cycle rickshaws, buffalo carts, sacred elephants on the way to some temple or other, animals of all kinds, and overloaded trucks, the two heavy loads and Pareek and Qureshi's white Ambassador car set out on the road to Bhopal. 'Every rut, every time a horn sounded, every acrobatic overtaking of a vehicle, every railway crossing, made us jump,' Shekil Qureshi remembered.

---

[1] Horse-drawn cart.

'Have you had any dealings with MIC before?' Pareek suddenly asked his companion, who was muttering prayers.

'Yes, once. A sparkling liquid in a bottle. It looked just like mineral water.' At this idea the two men broke into slightly strained laughter. 'In any case,' Qureshi went on, 'it was so clear, so transparent, you'd never have thought you had only to inhale a few drops for it to kill you.'

Pareek directed the driver to pass the two trucks and stop a little further on. The sun was so hot that he was worried.

'Our cans mustn't start to boil.'

The two men were well aware that methyl isocyanate started to boil as soon as it reached a temperature of 39° celsius. They also knew that the result could be catastrophic.

Qureshi put his head out of the window. A blast of burning air hit him in the face.

'I bet it's at least forty degrees, possibly even forty-five.'

Pareek grimaced and signalled to the driver of the front truck to stop. The two men at once rushed over to cover the drums with heavy isothermic tarpaulins. Then they took the extinguishers out of their holders. In case of danger, a jet of carbonic foam could lower the temperature of a drum by a few degrees.

'But we didn't harbour too many illusions,' the engineer admitted.

For thirty-eight hours, the two intrepid Carbide employees acted as sheepdogs, with their Ambassador sometimes in front of and sometimes following the two trucks. They had been given explicit instructions: their convoy must stop before entering any built-up areas to allow time to fetch a police escort. 'You could read the extreme curiosity on the local people's faces at the sight of these two trucks surrounded by police officers,' Pareek recalled. 'What can they possibly be transporting under their tarpaulins to justify that sort of protection?' people must have been asking themselves.

That first high-risk convoy was to be followed by dozens of others. Over the next six years, hundreds of thousands of gallons of the deadly liquid were to traverse the villages and countryside of Maharashtra and Madhya Pradesh. Until the day came in May 1980 when, to the euphoria of all the staff, the chemical reactors of Bhopal's brand new plant produced their first gallons of methyl isocyanate. These were immediately dispatched into three immense tanks capable of storing enough MIC to poison half the city.

❈

The city that had withstood invasions, sieges and the bloodiest of political plots, was in the throes of succumbing to the charms of a foreign chemical giant. Eduardo Muñoz could rejoice: Carbide was going to achieve by peaceful means what no one else had managed in three centuries: the conquering of Bhopal. In addition to the crescents on its mosques, the *linga* of its Hindu temples and the crosses on its church, the capital of Madhya Pradesh now paraded a secular emblem which was to forever alter its destiny: the blue and white logo of a pesticide plant. 'That prestigious symbol would contribute to the advent of a privileged class of workers,' Kamal Pareek explained. 'Whether you were employed at the very top of the hierarchy or as the humblest of operators, to work for Carbide was to belong to a caste apart. We were known as the "lords".' At Carbide, an engineer earned twice as much as an official in the Indian administration. This meant he could enjoy a house, a car, several servants, and travel in first class air-conditioned trains. What counted most, however, was the prestige of belonging to a universally recognized multinational. Social

119

status plays such a crucial a role in India. 'When people read on my business card: "Kamal Pareek – Union Carbide India Limited", all doors were opened,' the engineer recalled.

Everyone dreamed of having a family member or an acquaintance employed by the company. Those who had that good fortune were quick to sing its praises. 'Unlike Indian companies, Carbide did not dictate what you should do with your salary,' one of its managerial staff explained. 'It was American liberty overlaying an Indian environment.' For V.N. Singh, the son of an illiterate peasant from Uttar Pradesh, the envelope stamped with the blue and white logo that the postman delivered to him one morning 'was like a message from the god Krishna falling from the sky'. The letter inside informed the young mathematics graduate that Carbide was offering him a position as a trainee operator in its phosgene unit. The boy scrambled across the fields as fast as his legs would carry him to take the news to his father. His neighbours came running. Soon the entire village had formed a circle round the fortunate chosen one and his father. Both were too moved to utter a sound. Then a voice shouted: *'Union Carbide Ki Jai!* Long live Union Carbide!' All the villagers joined in the invocation, as if the entry of one of their own into the service of the American company were a benediction for all the occupants of the village.

As for the Muslim, Shekil Qureshi, who had taken part in the dangerous transportation of the drums from Bombay to Bhopal, joining Carbide as a trainee supervisor brought him a sumptuous marriage at the Taj ul-Masajid, the great mosque built by Shah Jahan Begum. Dressed in a glittering *sherwani*[2] of gilded brocade, his feet shod in slippers encrusted with precious stones, his arm entwined with the

---

[2] Long tunic.

traditional band inscribed with prayers soliciting the protection of Allah for him and his wife, with a red silk Rajasthani turban on his head, the young graduate in chemistry from Safia College proudly advanced towards the *mihrab*[3] of the mosque, 'dreaming of the linen boiler suit with the blue and white logo that was, as far as I was concerned, the finest possible attire.'

Such was the prestige conferred by a job with Carbide that families from all over came to Bhopal to find husbands for their daughters. One morning, sensing his end was near, Yusuf Bano, a cloth merchant in Kanpur, put his 18-year-old daughter Sajda on the express train to Bhopal with the secret intention of having her meet the son of a distant cousin, who was working in the phosgene unit on the Kali Grounds. 'My cousin, Mohammed Ashraf, was a handsome boy with a thick black moustache and a laughing mouth,' the girl recalled. 'I liked him at once. All his workmates and even the director of the factory came to our wedding. They gave us a very amusing present. My husband was moved to tears: two Union Carbide helmets with our forenames interlaced in gilded lettering.'

For the 26-year-old mechanical engineer Arvind Shrivastra, who was part of the first team recruited by Muñoz, 'Carbide wasn't just a place to work. It was a culture too. The theatrical evenings, the entertainment, the games, the family picnics beside the waters of the Narmada, were as important to the life of the company as the production of carbon monoxide or phosgene.'

The management constantly urged its workers to 'break up the monotonous routine of factory life', by creating cultural interest and recreational clubs. In an India where the humblest sweeper is reared on tales from history and

[3] Niche indicating the direction of Mecca – and therefore of prayer.

121

mythological epics, the result exceeded all aspirations. The play entitled *Shikari ki bivi* put on by the workers from the phosgene unit was a triumph. It exalted the courage of a hunter who sacrificed himself to kill a man-eating tiger. As for the first poetry festival organized by the Muslims working in the 'formulation' unit, it attracted so many participants that the performance had to be extended for three further nights. Then came a magazine. In it the operator of the carbon monoxide unit, who was also the editor-in-chief, called upon all employees to send him articles, news items and poems, in short any material that might 'bring in ingenious ideas to contribute to everyone's happiness'.

These initiatives, which were typically American in inspiration, soon permeated the city itself. The inhabitants of Bhopal may not have understood the function of the chimneys, tanks and pipework they saw under construction, but they all came rushing to the cricket and volleyball matches sponsored by the new factory. Carbide had even set up a highly successful hockey team. As a tribute to the particular family of pesticides to which Sevin belonged, it called its players 'the Carbamates'. Nor did Carbide forget the most poverty-stricken. At Diwali, young Padmini saw an official delegation sent by the company to the Orya Bustee, with baskets full of sweets, bars of chocolate and biscuits. While the children launched themselves at the sweets, other employees went round the huts, distributing what Carbide considered to be the most useful gift in overpopulated India: condoms.

❁

As for the Americans, who through their company's development in Bhopal had been catapulted into the heartland of India, they felt as if they had landed on another planet.

In the space of twenty-four hours, 44-year-old Warren Woomer and his wife Betty had gone from their peaceful, aseptic West Virginia to the bewildering maelstrom of noises, smells and frenetic activity of the City of the Begums. For this highly skilled employee, to whom the company would shortly entrust the command of the Bhopal factory, the adventure was 'a real culture shock'. 'I knew so little about India!' he candidly admitted. 'I realized we'd have to adjust our thought processes and way of life to thousand-year-old traditions. How were we going to get a bearded, turbaned Sikh to put on a mask when performing dangerous procedures? Before I left Charleston, I didn't even know what a Sikh was!' For his young compatriot, John Luke Couvaras, who, in his enthusiasm, had likened the Bhopal venture to 'a crusade', 'the experience was absolutely unique'. 'I particularly remember the feeling of excitement,' he said, 'but India never failed to endear itself to us, sometimes quite comically.'

In the beginning, employees regularly arrived late to their workstations.

'*Sahib*, the buffalo cows had escaped,' one of Couvaras' workers excused himself one morning. 'I had to run after them to milk them.'

The American admonished the former peasant gently.

'The running of our factory cannot depend on the whims of your buffalos,' he explained.

'After six months, everything was working to order,' Couvaras said.

Plenty of other surprises would strike the young engineer, starting with the perceived difference in mentality between Hindu and Muslim engineers. 'If there was a problem, a Muslim would give you the facts straight before acknowledging his responsibility. Whereas a Hindu would remain vague and then incriminate fate. We had to adapt ourselves

123

to these differences. Fortunately, after a certain level of education, the chemistry fairy intervened to put us all, Indians and Americans alike, on the same wavelength.'

# 19

## *The Lazy Poets' Circle*

'**M**y very dear Engineer Young, your presence does us infinite honour. Be so good as to remove your shoes and stretch out on these cushions. Our poetry recital is due to commence in a few moments. While you're waiting, do quench your thirst with this coconut.'

Thirty-one-year-old Hugo Young, a mechanical engineer originally from Denver, Colorado, could scarcely believe his eyes. He had suddenly found himself thousands of light years away from his phosgene reactors, in the vast drawing room of one of Bhopal's numerous patrician residences. About him, some twenty men of different ages were reclining on silk cushions embroidered with gold and silver, their heads resting on small brocade pillows, the property of each guest. By buying these pillows they had acquired the right of entry into the City of the Begums' most exclusive club, the 'Lazy Poets' Circle'. Bhopal might be being precipitated into the industrial era, but as one expatriate from the Kanawha Valley witnessed, it was not going to give up any of its traditions in the process. All the adepts of the Lazy Poets' Circle continued to observe the very particular laws and rites of their brotherhood. Those 'reclining' were considered to be lazy poets of

125

the first order; those 'seated', lazy poets of the second order; as for those standing, they were voluntarily depriving themselves of the respect of their peers. This hierarchy of posture entitled the 'reclining' to command the 'seated' and the seated to command the 'standing'. It was a subtle philosophy, which even found its expression in material things. Thus cups and bowls with thick rims were strictly prohibited to save members of the Lazy Poets' Circle from having to open their lips any wider than necessary when drinking.

All afternoon, poets, singers and musicians succeeded one another at the bedsides of the 'lazy', charming them with their couplets and aubades. In the evening, after an army of turbaned servants had served them all kinds of samosas, the brotherhood took the young American to the parade ground in the old town where a poetry festival was being held. That evening, the *mushaira* had brought together several authors, professional and amateur, who were singing their works to a particularly enthusiastic audience. 'My friends made a point of translating the *ghazals*[1] for me,' Young remembered. 'They all evoked tragic destinies, which love saved in the end. As I listened to the voices with their harmonies rising ever higher until they sounded almost like cries for help, I thought with embarrassment of the fatal phosgene I was making in my reactors only a few hundred yards from that prodigious happening.'

In the course of the evening one of the members of the Poets' Circle placed a hand on the young American's shoulder.

'Do you know, dear Engineer Young, which is the most popular *mushaira* in Bhopal?' he enquired.

The engineer made a show of thinking about it. Then with a mischievous wink, he replied:

---

[1] Poetic couplet.

'The Lazy Poets', I imagine.'

'You're way off, my dear fellow. It's the *mushaira* of the municipal police. The chief of police told a journalist one day that it was "better to make people cry through the magic of poetry than with tear gas".'

❅

Indolent, voluptuous, mischievous and always surprising – that was Bhopal. John Luke Couvaras would never forget the spectacle he came across one afternoon in the living room of his villa in Arera Colony. Stretched out on a divan, his young Canadian wife was being massaged by two exotic creatures with kohl-rimmed eyes and heavy black tresses that tumbled to their thighs. The grace of their movements, their delicacy and concentration extracted a string of compliments from the engineer, but the thanks he received in response could have come from the mouths of a pair of Liverpool dockers: the long hands decorated with geometric patterns in henna, kneading away at his wife's flesh, belonged to two eunuchs.

Less than 800 yards from the futuristic complex rising from the Kali Grounds, in old houses washed out by the monsoon, lived a whole community of *hijras*, a very particular caste in Indian society. They had come to the City of the Begums from every region of India for festivals and pilgrimages and then stayed. In Bhopal, three or four hundred eunuchs were reckoned to be established in small groups around a guru who acted as head of the family. Apart from having a talent for massage, they played an important role in local Hindu society. To these beings, neither men nor women, religion attributed the power to expunge sins committed by new-born babies in their previous lives. Whenever there was a birth, they came running with their tambourines coated in red powder for the ceremony of purification. They were always generously remunerated. No one in Bhopal would haggle over the services of the *hijras* for fear of incurring their maledictions.

❋

The South Charleston expatriates 'culture shock' arose from experiences of a kind only India could offer. For the 36-year-old bachelor, Jack Briley, an alpha naphtol expert, the Orient and its charms were embodied as one woman. She was one of the Nawab's nieces: he had met her at a cocktail party in honour of the President of the World Bank. Refined, cultured and liberated, gifted moreover with a lively sense of humour, Selma Jehan was, with her large eyes highlighted with kohl, 'the perfect incarnation of a princess out of *A Thousand and One Nights* of the kind a young American from the banks of the Kanawha might dream of.' Jack Briley allowed himself to fall easily under her spell. As soon as he could escape from the plant, the young Muslim girl showed him the city of her ancestors. As the rules of *purdah*[2] ordained, the windows of the old family Ambassador, which she drove herself, were hung with curtains to hide her passengers from others' sight.

Selma took her suitor to the city palaces where some members of her family were still living. Most of these buildings were in a sorry state with cracked walls, ceilings occupied by bats and grimy furniture.

Other residences housed the survivors of another age. Zia Begum, Selma's grandmother, lived amongst her bougainvillaeas, and her neem and tamarind trees on Shamla Hill.

---

[2] Muslim law which obliges women to conceal their faces and bodies from the eyes of men.

She never failed to show visitors the silver-framed portrait of the first gift she had received from her husband: a 16-year-old Abyssinian slave in Turkish trousers with a waistcoat embroidered with gold.

Briley had the good fortune to be a guest at several receptions held by this unusual grandmother. There he met all the town's upper crust, people like Dr Zahir ul-Islam, who had just successfully performed Bhopal's first sex-change operation, or the little man they called the Pasha, the town's gossip, full of tittle-tattle. Wearing a wine-coloured fez and a suit of silver embroidered brocade, his eyes made up with kohl, the Pasha spoke English with an Oxford accent. He had lived in England for twenty years but left because he said he felt too Indian there. He found living in India difficult, because he felt too English. Only in Bhopal did he feel at home.

Another frequenter of Zia's soirées was an eccentric old man in rags, known as Enamia. Under his real name of Sahibzada Sikander Mohammed Khan Taj, this obscure, impecunious cousin of the Begum had married a Spanish princess. He too had spent twenty years in London where he worked in a sausage factory before getting himself dismissed for unhygienic behaviour. No one had ever tried to find out what lay behind those words, but the Begum and her friends doted upon old Enamia. He was one of the great connoisseurs of the city and nothing gave him greater pleasure than taking foreign visitors round it in his old jeep with its tired shock absorbers. He knew the history of every street, monument and house. Enamia was Bhopal's memory.

The Begum's dinners also brought together passing artists, politicians, writers and poets. One of the regulars was of course Eduardo Muñoz, to whom Bhopal owed the arrival of Carbide. It was unanimously agreed that one of the major attractions of these dinners was the excellence of the food.

The best in Bhopal – so it was claimed. For young Briley every invitation was a gastronomic experience. It was there that, for the first time in his life, he tasted partridge cooked in coriander and sweets made out of curdled milk in a syrup of cinnamon and ginger.

It had become a tradition: the weddings of the Begum's grandchildren, nephews and nieces were always held at her home under an immense *shamiana*[3] erected in the courtyard. They were the occasion for three days of uninterrupted celebrations. The drawing rooms, courtyards and corridors of the palace were littered with divans on which guests reclined to drink and listen to *ghazals* and other poetic forms. Selma, the woman whom Briley loved, had grown up performing every kind of Hindu dance. She would decorate her ankles and wrists with strings of bells, then appear on the dais and give passionate performances of *kathak*, a southern Indian dance accompanied by the complex rhythms of *tabla* and *sarod* players. Then the scent of patchouli and musk floating beneath the *shamiana* would become so intoxicating that the American thought he would never again be able to put up with the smell of phosgene or MIC.

※

Not all the expatriates from South Charleston in the City of the Begums were lucky enough to have a love affair with a princess. But the attractions of Bhopal were numerous, starting with the uninterrupted succession of religious festivals, celebrations and ceremonies. There was the *bujaria*, the noisy, colourful procession of thousands of eunuchs, the *hijras*, that went all round the old town; and the great Hindu festival in honour of the goddess Durga, whose richly

---

[3] Large tent for festivities and ceremonies.

decorated statues were immersed in the lake in the presence of tens of thousands of faithful. Then there was the Sikh celebration of the birth of Guru Nanak, the founder of their religion, with firecracker explosions that woke up the whole town. And there was the Jain festival in honour of their prophet Mahavira and the return of the pilgrimage season marked by the official end of the monsoon. Autumn brought Eid and Isthema, two Muslim festivals that drew hundreds of thousands of followers to the old part of town, as well as the many other religious and secular celebrations that reflected the extraordinary diversity of the people of Bhopal.

# 20

## 'Carbide has poisoned our water!'

One was called Parvati, after the wife of the god Shiva; another Surabhi, 'the cow with all gifts', born, according to the *Vedas*, of the great churning of the sea of milk; a third was Gauri the light; and the last two were Sita and Kamadhenu. So gentle were they that little children were not afraid to stroke their foreheads above the large eyes surrounded by lashes so long they looked as if they were wearing make-up. These five cows were some of the 300 million head that made up the world's premier cattle stock. For the five families in Orya Bustee to whom they belonged, they were an enviable asset. Belram Mukkadam, Rahul, Padmini's father, Ganga Ram and Iqbal were the lucky owners of this modest herd. The few pints of milk they gave each day provided a little butter and yoghurt, the only animal protein available to the hungry people of the bustee apart from goat's milk. Religiously collected up and made into cakes, the dung from these cows was dried in the sun on the walls of the huts and used as fuel for cooking. Each animal knew its way home and returned to its owner in the evening after roaming about all day in search of a little greenery on the edges of the Kali Grounds. On the twelfth day of Asvin's moon, in

September, of Kartika's in November and during the festival of the new rice, the owners dyed the cows' horns blue and red and decorated them with garlands of marigolds and jasmine. The animals were arranged in a semicircle outside Belram Mukkadam's tea house, so the sorcerer Nilamber could recite mantras over them. As the neighbourhood's most long-standing resident, it fell to Mukkadam to make the customary speech.

He did so with particular feeling: 'Each one of our cows is a celestial animal, a symbol of the mother who gives her milk,' he declared. 'She was created on the same day as Brahma, founder of our universe, and every part of her body is inhabited by a god, from the nostrils where Asvin dwells to the fringing of her tail, where Yama resides.' The sorcerer Nilamber, in his saffron robe, intervened in his turn to emphasize 'how sacred everything that comes from the cow is'. Upon these words, Rahul brought a bowl filled with paste. It was the traditional purée made out of gifts from the precious animal – milk, butter, yoghurt, dung and urine. The receptacle was passed from hand to hand so that everyone could take a small ball of the purifying substance. Led by Padmini, young girls then spread a little earth and fresh dung mixed with urine over the mud flooring of the huts. This protective layer had the power to repel scorpions, cockroaches and above all, mosquitoes, the persistent scourge of the Bhopalis.

That autumn festival day, Mukkadam had a special mission of his own. As soon as the ceremony was over, he attached a garland of flowers to the horns of his cow and led her away to his hut at the end of the first alleyway. Inside the one and only room, the elderly father of the manager of the tea house lay stretched out on a *charpoy*, watched over by his two daughters who fanned him and uttered prayers. His halting breath and dull eyes suggested that death was imminent. Mukkadam pushed the cow over to the dying man's bedside, then took the tip of her tail and tied it with a piece of cord to his father's hand.

'Lead this holy man from the unreal to the real, from darkness to light, from death to immortality,' he murmured, as he stroked the animal's forehead gently.

✼

Four days after the death of Belram Mukkadam's father, a catastrophe befell the inhabitants of Orya Bustee. Padmini was drawing a bucket of water from the well when a noxious smell issued from the shaft. The water was a strange whitish colour. The old woman Prema Bai plunged her hand into the bucket to scoop up a little of the liquid and taste it.

'This water is contaminated!' she announced with a grimace of disgust.

All the other women present confirmed her verdict. Looking up at the steel structures that loomed on the horizon, Padmini's mother shouted:

'Come on everyone! Come and see! Carbide has poisoned our water!'

✼

A few hours later, Rahul and several of the neighbourhood's young men burst into the tea house.

'Belram, come quickly!' cried the cripple. 'Your cow Parvati and all the other cows are dead. The crows and vultures that consumed their corpses are dead too.'

Mukkadam set off at a run for the place the boys had indicated. The animals lay stretched out beside a pool fed by a rubber pipe issuing from the factory. 'It's water from

Carbide that's killed them,' he said angrily. 'The same water that has poisoned our well. Let's all go to Carbide, quickly!'

A cortege of 300 or 400 people promptly set off on a march to the factory. Omar Pasha and his sons, Ganga Ram, the shoemaker Iqbal, his friend Bassi the tailor, and the bicycle repairman Salar marched at the head. Even the dairyman Bablubhai and Nilamber went. 'Pay us compensation for the cows! Stop poisoning our well!' they yelled in chorus. In the second row came six men, bent beneath the weight of a *charpoy* they were carrying on their shoulders. On this string bed they had placed the hide of the multinational's first victim in Bhopal. The horns painted with sandalwood paste, visible between the folds of the shroud, revealed that it was a cow. 'Today it's our cows. Tomorrow it will be us!' ranted the angriest members of the cortege. Hope of employment and the prestige of the uniform with the Carbide logo continued to feature in people's dreams, but these deaths shattered the illusion of living in neighbourly harmony.

The plant management appointed one of the engineers to settle the matter as quickly as possible. The American stood in front of the demonstrators.

'Friends, set your minds at rest!' he shouted into the megaphone, 'Union Carbide will compensate you generously for your loss. If the owners of the cows that have died will just put up their hands!' The engineer was astonished to see a forest of hands promptly raised aloft. He took a bundle of notes out of his pocket. 'Union Carbide is offering five thousand rupees for the loss of each animal,' he announced. 'That's more than ten times the price of each of your cattle. Here are twenty-five thousand rupees. Share them between you!'

He held out the wad of notes to Mukkadam.

'And the water in our well?' insisted Ganga Ram.

'Don't worry. We'll have it analysed and take whatever steps are necessary.'

The results of the tests were so horrific that the factory management stopped them being divulged. Samples of soil taken from outside the periphery of the Sevin formulation unit revealed high levels of mercury, chromium, copper, nickel and lead. Chloroform, carbon tetrachloride and benzene were detected in the water from the wells to the south and south east of the factory. The experts' report was explicit: this was a case of potentially deadly contamination. Yet, for all the promises of Carbide's representative, nothing was done to stop the pollution.

❉

The envelope bore the stamp of the Indian Revenue Service. It contained the government's official tribute to the man who for nine years had been fighting to provide Indian agriculture with the means to defend itself against the microscopic hordes that ravaged its crops. Eduardo Muñoz started when he read the letter inside the envelope. Becoming a fully paid-up tax payer in the Indian Republic was not exactly one of his greatest aspirations, especially when, as the fiscal services informed him, he owed almost 100 per cent tax on his salary. He decided to pack his bags.

'Leaving India after all those thrilling years was heartbreaking,' the father of the Bhopal factory acknowledged. 'But I left feeling confident. The Indian government had confirmed that Carbide was authorized to make all the ingredients for the production of Sevin on the Bhopal site. The document was numbered C/11/409/75. After a long and difficult struggle, my "beautiful plant" was soon, in the words of our advertising slogan, to bring the people of India, "the promise of a bright future".'

Muñoz's optimism was, at very least, ill founded. He was probably not aware that the people of the Kali Grounds' bustees had made their first stand against the harmful effects of his 'beautiful plant'. The state of the country he was leaving was even more worrying. India was once again suffering from drought. All through the month of June, millions of men, women and children had watched the sky for the first signs of the monsoon. Usually a buffeting wind gets up a few days before the first storm breaks. Suddenly the sky darkens. Huge clouds roll in upon each other, scudding along at a fantastic rate. Other clouds succeed them, enormous, as if trimmed with gold. A few moments later, a mighty gust of wind brings a hurricane of dust. Finally a new bank of black clouds plunges the sky into darkness, an interminable roll of thunder rends the air, and the monsoon has begun. Agni, the fire god of the *Vedas*, protector of humanity and their hearths, hurls his thunderbolts. The great warm drops turn into cataracts of water. Children throw themselves, stark naked, shrieking for joy, into the deluge. Men are exultant and, under the verandas, women sing hymns of thanksgiving.

That year, however, in several regions, water, life and rebirth failed to keep their appointment. Their seedlings parched, in the stranglehold of debt, millions of ruined peasants had been unable to buy fertilizers or pesticides. In 1976 the sales figures for Sevin had dropped by half. Another severe blow after the drought of the previous year.

Nonetheless a pleasant surprise awaited Eduardo Muñoz on his return to New York. In recognition of his faithful services, the company had appointed him President of the International Division of its agricultural products. The installation ceremony took place at Carbide's new head office. They had recently sold their Park Avenue property to Manufacturer's Hanover Trust Bank, a move that so distressed the municipal government that the Governor of New York, Hugh Carey, and two senators had tried to dissuade Bill Sneath from moving the prestigious multinational out of Manhattan. They had offered him subsidies and tax advantages. In ten years the city had lost the head offices of forty-four of the largest American companies, a fact which had meant the loss of some 500,000 jobs. All the promises in the world could not persuade Carbide's Chairman to change his mind. He had systematically enumerated the disadvantages of New York, a city which both he and his colleagues judged to be overpopulated, expensive and unsafe, where the standards of education were execrable, transport was lacking, and taxes were exorbitant. The company had designs on a particularly imposing site set in the middle of a hundred-acre estate, which was home to deer and bucks. It was situated near Danbury, a charming little town in Connecticut with a hat factory that had been supplying sheriffs, senators, gangsters and America's middle classes for two centuries. The new headquarters were shaped in the form of an airport terminal with satellite wings, underground parking, auditoriums, lecture rooms, libraries, a bank, five restaurants, a fitness centre, hospital, hair-dresser's, gift-shop, newspaper kiosk, travel agent and car-hire centre, a television studio, a printer's, an information centre, acres of air-conditioned offices, and even a one and a quarter mile jogging track. All the evidence suggested that the proud manufacturers of methyl isocyanate had found a headquarters to suit the company's renown, its importance and its ambitions for the planet. It was said to have cost a trifling 800 million dollars.

❈

In the peaceful suburbs of West Virginia, in the vicinity of the Institute 2 industrial site, the smell was unfamiliar. It was not MIC's boiled cabbage, but the aroma of the small, fiery, red chillies that enhance the flavour of spicy Indian cooking. 'They rustled up their food in the rooms we'd rented for them,' Warren Woomer explained. On his return from India, he had been assigned to supervise the twenty or so Indian technicians and engineers sent over by the Bhopal factory. At the end of 1978, they were undergoing a six-month long intensive training period in the various units of the American plant. Woomer remembered the amazement of the enthusiastic group as they discovered America. 'The Indian government had only authorized them to bring out five hundred dollars per person, but you can't begin to imagine what an Indian can do with five hundred dollars! In the evenings and at weekends they would descend upon camera or radio shops like locusts and set about haggling oriental-style, extracting astronomical reductions that we Americans would never have managed to get.'

But the Bhopal trainees had not come half way round the world to shop. For each one of them Woomer had prepared a rigorous work programme designed to prepare them for the imminent launch of their factory. 'It was an invaluable experience,' young Kamal Pareek, who was on the trip, said later, 'even if our factory was only a child's toy compared with the Institute monster that, day and night, went on producing seven times more Sevin than ours would ever make.' Realizing that a ship of 100 tons poses the same navigational and maintenance problems as a 50,000 ton battleship, Woomer assigned each visitor to the department dealing with his speciality, whether it was handling gases, working the reactors, operating the electrical circuits and control systems, producing MIC or maintaining and repairing the installations; manufacturing phosgene, 'formulating' Sevin, preventing corrosion, or indeed gestating toxic waste, protecting the environment, or running the company. With

139

on-site instruction sessions, audio-visual shows, training periods in laboratories, and visits to the suppliers and manufacturers of equipment, Woomer and his team made every effort to bring about what the American called 'an appropriate transfer of knowledge'. Each visitor was instructed to put what he was learning down in writing so that on his return to Bhopal he would be able to compile an instruction manual for his fellow employees.

One of the most significant 'transfers of knowledge' was not of a technical nature; it was a message of a rather different order. In a curious doctrine, combining cynicism with realism, the company's managers had defined the principles of a methodology they called *corporate safety. Human beings are our most precious asset*, affirmed the preamble to the doctrine's manifesto, *and their health and safety are therefore our number one priority*. Carbide might have been expected to feel some scruples about delivering such a message even as its factories were contaminating the very valley in which now it was receiving its Indian visitors. It did not.

'How could we not enthusiastically applaud such a profession of faith,' Pareek would ask, 'when we were responsible for assuring the safety of the first plant to produce methyl isocyanate outside America?' Carbide's manifesto set down certain truths, the first being that *all accidents are avoidable provided the measures necessary to avoid them are defined and implemented*. But it was on another more subtle argument that the multinational's management depended to impress upon their visitors the importance of safety. The formula they came up with was simple: *Good safety and good accident prevention practices are good business.*

'At Institute, Union Carbide's real emblem was not the logo but a green triangle inscribed with the words "SAFETY FIRST"' Kamal Pareek, the future assistant manager for safety at the Bhopal factory, stated in naïve admiration. This obsession with safety manifested itself primarily through the study of a voluminous 400-page manual outlining in minute detail the instructions for emergency procedures to be carried out in case of accident, on how to keep personnel continuously informed, on the constant checking of all apparatus, regular practices for safety crews and equipment, as well as the immediate identification of toxic agents, evacuation procedures and a thousand other extreme situations. 'At Institute, the Indian engineer explained, 'the posters the management seemed most proud of were not graphs tracking the rising curve of Sevin sales, but the safety awards the company's various factories throughout the world had won.'

# 21

## The first deadly drops from the 'beautiful plant'

No plaque commemorates the launch of the *Titanic*. Nor does any history book make reference to 4 May 1980, the date that the first plant exported from the West to make pesticides using methyl isocyanate, began production. Yet for the men who had built it, that day was 'cause for jubilation' as one of them would later say. Thirteen years after Eduardo Muñoz's grey Jaguar had first rolled up at the Kali Grounds, a dream was coming true.

With speeches, the handing out of gifts, garlands and sweets, the company had assembled several hundred guests under multicoloured *shamianas* to mark the occasion. Official dignitaries, ministers, senior civil servants, directors of the company, personnel from the various units, ranging from the foreman to the humblest operator, stood at the foot of this remarkable structure. The engineers, American and Indian, made no secret of their delight and relief at having surmounted the obstacles of a long and difficult process.

The new Chairman of Union Carbide, Warren Anderson, had come over especially from the United States for the event. Tall, athletic-looking, with a white plastic safety helmet on his thick grey hair, Anderson towered above the assembly. The son of a humble Swedish joiner who had immigrated to Brooklyn, at fifty-nine he epitomized the fulfilment of the American dream. Equipped with a degree in chemistry and another in law, in thirty-five years he had climbed the ladder to the top of the world's third largest giant of the chemical industry. The empire he now ran comprised 700 plants employing 117,000 people in thirty-eight countries. For this passionate line fisherman who loved gardening at his home in Connecticut, the birth of the new plant was a decisive step towards his life's principal objective. Anderson wanted to turn Union Carbide into a company with a human face, a firm in which respect for moral values would carry as much weight as the rise of its shares on the stock market. Thanks to the Sevin that the Carbide teams were going to manufacture here, tens of thousands of peasants could protect their families from starvation. With a garland of marigolds round his neck, Chairman Warren Anderson had every reason to be proud and happy. This plant was his triumph.

❄

Getting the installation up and running had involved three challenging months of intensive preparation. Finding and training technicians who could confront any eventuality, right in the heart of India, had been no easy matter. The list of possible incidents and the problems they were likely to engender had no fewer than eighty, sometimes extremely serious, entries. 'You don't launch such a complex plant like you turn the ignition key in a car,' Pareek explained. 'We were dealing with a kind of metal dinosaur, complete with its bad temper, its whims, its weaknesses and its deformities. Waking a monster like that up and bringing it to life,

143

with its hundreds of miles of piping, its thousands of valves, joints, pumps, reactors, tanks and instruments was a task for the pharaohs.'

The task began with a rigorous check of the sealing of all circuits by flushing all the pipework repeatedly with nitrogen. To detect any leaks in the connecting joints, safety valves, pressure gauges and sluices were smeared with a soapy coating. The smallest bubble alerted the operators. Next, one by one, all the hundreds of bolts that held together the various pieces of equipment had to be tightened. Once the system had been found to be functioning correctly, the engineers began heating up the two gases which, when brought together, would produce methyl isocyanate. These two components – phosgene and monomethylamine – had themselves been obtained by combining other substances. As the temperature of the gases rose, the operators opened up the circuits one by one. The few privileged people present in the control room held their breath. The fateful moment was approaching. John Luke Couvaras checked the dials on the reactors' temperature and pressure gauges. Then he cried: 'GO!' Whereupon an operator activated a circuit that sent the phosgene and the monomethylamine into the same steel cylinder. The combination produced a gaseous reaction. This gas was at once cooled down again, purified and liquefied. Then came a burst of applause. Six years after setting off an atomic explosion, India had just produced its first drops of methyl isocyanate.

'We weren't able to see the first trickle of MIC,' Pareek recalled, 'because it went straight into the holding chamber. But as soon as the chamber was full we put on our protective suits to take a sample of a few centilitres of the liquid. I carried the container with as much respect as if it had been a statue of Durga, into the laboratory to have the contents analysed. We were thrilled at the result. Our Indian MIC was as pure a vintage as Kanawha Valley's!'

❆

While Union Carbide's tanks filled up to the accompaniment of general euphoria, on the southern boundary of the Kali Grounds a celebration of a very different kind was going on. Belram Mukkadam, Rahul, Ganga Ram, Ratna Nadar and many of the other residents of Orya Bustee had gathered round five horned beasts just delivered by a cattle merchant. With the indemnity money paid out by Carbide, Mukkadam had decided to replace his cow Parvati with a bull. He called it Nandi, after the bull the god Shiva rode, because Nandi kept all danger and evil at bay. That night, by the light of the full moon, he marked the animal's forehead with the trident of the god. It was an emblem that augured well, Mukkadam was sure of it: it would guarantee the fertility of the new herd and ensure divine protection of the Kali Grounds' bustees.

# 22

## *Three tanks dressed up for a carnival*

By appointing one of its best men to the board of its Indian pesticide plant, the American multinational was signalling the degree of control it expected to exercise over the Bhopal development. Modest, almost timid-looking behind his thick glasses, Warren Woomer was one of Carbide's most experienced and respected engineers. Neither India nor Bhopal were unknown to him. Woomer had carried out two assignments there. One had been to help his Indian colleagues get their unit producing alpha naphtol, a substance used in the composition of Sevin, up and running. The other had been to assist the same Indians in the launching of their plant and check that they were applying everything he had taught them in Institute correctly.

Being an American in charge of about a thousand Indians of different origins, castes, religions and languages was the toughest challenge of his career. Woomer began with a detailed inspection of the ship: 'I couldn't find anything fundamental to fault. Of course the control room would seem obsolete to us now, but at the time it was the best that India could produce. I noticed nothing really shocking about either the design or the functioning of the plant. In any case my bible was the *MIC: manual of use* with its forty pages of instructions. Every one of them was to be treated as Gospel truth, especially the directive to keep the MIC in the storage tanks at a temperature close to zero degrees celsius. On this point I had decided to be intractable. Yes, it was imperative that every single drop of MIC was kept at zero degrees. What's more, my long honeymoon with some of the most dangerous chemical substances made me add a recommendation to the *MIC: manual of use*. I considered it vitally important: only stock a minimum quantity of methyl isocyanate on site.'

Although he had encountered no problems at a technical level, Woomer still realized that many things could be improved, notably the way in which staff members performed their tasks: 'For example, no one took the precaution of wearing safety goggles,' he remembered. 'One day I put my hand over one of the operator's eyes. "That's how your children and grandchildren are likely to see your face if you don't protect your eyes," I told him severely. The story did the rounds of the plant and, next day, I found everyone wearing safety goggles. I realized then that in India you had to touch people through the heart.'

There were plenty of other problems in store for the new captain. Firstly, how was he to remember the unpronounceable names of so many of his colleagues?

'Sathi,' he said one day to his secretary, 'you're going to teach me the correct pronunciation of the first and last names of everyone working in the plant, including those of their wives and children. And I'd like you to point out any mistakes I make because of my ignorance of the ways and customs of your country.'

'*Sahb*[1], in India, employees don't tell their bosses what to do,' responded the young Indian girl.

---

[1] Affectionate abbreviation of 'Sahib'.

'I'm not asking you to tell me what to do,' replied Woomer sharply. 'I'm asking you to help me be as good a boss as possible.'

As good a boss as possible! Warren Woomer was to discover, often at cost, the extreme subtlety of relationships in Indian society where every individual occupies a special place in a myriad of hierarchies. 'I learned never to make a remark to anyone in the presence of his superior,' he said later. 'I learned never to announce a decision without everyone having had the chance to express a view so that it appeared to be the result of a collective choice. But, above all, I learned who Rama was, who Ganesh, Vishnu and Shiva were; what events the festivals of Moharam or Ishtema commemorated; who Guru Nanak was and who was the God of Work my workers worshipped so ardently and whose name was so difficult to remember.'

※

Warren Woomer could not remain ignorant of Vishvakarma, one of the principal giants in the Hindu pantheon. In Indian mythology he personifies creative power. The sacred texts glorify him as the 'architect of the universe, the all-seeing god who disposes of all the worlds, gives the divinities their names and exists beyond mortal comprehension'. He is also the one who fashions the weapons and tools of the gods. He is lord of the arts and carpenter of the cosmos, builder of the celestial chariots and creator of all ornaments. That is why he is the tutelary divinity of artisans and patron of all the crafts that enable humankind to subsist.

Every year after the September moon, his effigy is borne triumphantly into all workplaces – from the smallest workroom to giant factories. This is a privileged time of communion between bosses and workers, when celebrations

148

unite rich and poor in shared worship and prayer.

Overnight the reactors, pumps and distillation columns of the 'beautiful plant' were decorated with wreaths of intertwining jasmine and marigolds in honour of Vishvakarma. The three great tanks due to contain tens of thousands of gallons of MIC were draped in fabrics of many colours, making them look like carnival floats. The vast Sevin 'formulation' unit, where the festivities were to be held, was covered in carpets and its walls were decorated with streamers and garlands of flowers. Some workmen brought cases full of hammers, nails, pliers and hundreds of other tools, which they deposited on the ground and decorated with foliage and flowers. Others set up a colossal altar in which the image of the god would be installed on a cushion of rose petals. Riding on his elephant covered with a cloth encrusted with precious stones, the statue resembled that of a maharajah. Vishvakarma wore a tunic embroidered with gold thread and studded with jewels. One could tell he was not a human being in that he had wings and four arms brandishing an axe, a hammer, a bow and the arm of a balance. Several hundred engineers, machine operators, foremen and workmen, most accompanied by their wives and children, all dressed up in their festival clothes, soon filled the work floor. Squatting, barefoot in the midst of this sea of humanity, Warren and Betty Woomer, the only foreigners, watched the colourful ceremony with astonishment and respect.

After intoning mantras into a microphone, a pundit with a shaven head placed on a *thali*, a ritual silver plate, the various appurtenances of the ceremony: first the purificatory fire – burning oil in a clay dish – then rose petals, a few small balls of sweet pastry, a handful of rice and finally the *sindoor*, a little pile of scarlet powder. Ringing his small bell vigorously, the pundit blessed the collection of tools laid out by the workers. A solitary voice then rang out, promptly followed by a hundred others. *Vishvakarma Ki Jai!* 'Long live Vishvakarma!' That was the signal. The ceremony was over but the festivities could commence. The management of the factory had arranged for a banquet of meat curry and vegetables, lassi, and *puri*[2] to be prepared in a nearby kitchen. Beer and palm wine flowed like water. The alarm system's loudspeakers poured out a flood of popular tunes and firecrackers went off on all sides. Employers and employees gave themselves up to celebration.

⁂

Like most of those responsible for running the 'beautiful plant', Warren and Betty Woomer were not aware that the occupants of the neighbouring bustees were gathered with similar fervour round the God of Tools. There was, after all, an extraordinary concentration of workers in those areas too. The workshops belonging to the shoemaker Iqbal, the sari-embroiderer Ahmed Bassi and the bicycle repairman Salar, were just three small links in a whole chain of work-places, in which devotees of Vishvakarma laboured in order to survive. In Jai Prakash and Chola, children supported their families by cutting up sheets of brass to make tools, or dipping fountain pen caps in chrome baths, which gave off noxious fumes. Elsewhere youngsters slowly poisoned themselves, making matches and firecrackers, handling phosphorous, zinc oxide and asbestos powder. In the poorly ventilated workshops, emaciated men laminated, soldered, and fitted pieces of iron work together, amidst a smell of burning oil and overheated metal. A few paces away from the spacious house belonging to the Sikh usurer Pulpul Singh, a

---

[2] Little cakes of fried corn puffed up into balloons.

dozen men, sitting cross-legged, made *bidis*. They were nearly all tuberculosis sufferers who did not have the strength to pedal a rickshaw or pull a *tilagari*[3] . Provided they did not stop for a single minute, they could roll up to 1300 cigarettes a day. Every evening a *tharagar* would come from the town to collect what they had produced. For 1000 *bidis*, they received 12 rupees, the price of 2 kilos of rice.

How surprised Chairman Anderson and Warren Woomer would have been if ever they had chanced upon those places where so many men and children spent their lives making springs, truck parts, axles for weaving looms, bolts, vehicle petrol tanks and even turbine gears to the tenth of a micron; men and children, who with a surprising degree of dexterity, inventiveness and resourcefulness, could produce, copy, repair or renovate any part or a machine. Here the least scrap of metal, the lowliest bit of debris was reused, transformed, adapted. Here nothing was ever scrapped. Everything was always reborn as if by some miracle.

In anticipation of the festival, labour had stopped in the workshops on the previous day, and everyone had scrambled to clean, repaint and decorate them with garlands of foliage and flowers. The workers of Orya Bustee, Chola and Jai Prakash did the God of Tools proud also.

In the space of one night, hell-holes had been transformed into places of worship adorned with temporary altars sumptuously decorated and strewn with flowers. The traditional chromo of the four-armed god perched on his elephant presided everywhere. Yesterday's slaves had changed into gleaming shirts and brand new *lunghis*; their wives had got out their festival saris, preserved in the family chests from the greed of the cockroaches. The children were equally resplendent. The entire local population squeezed in behind

---

[3] Hand cart.

a brass band, the flourishes of which resounded through the alleyways. The godfather Omar Pasha was present, with his two wives on either side of him, dressed up like queens in silk saris that Ahmed Bassi had embroidered and encrusted with pearls. The Muslim tailor was there, for the festivities transcended all religious differences. With his crony, the goateed mullah beside him, the sorcerer Nilamber, who was acting as pundit, led the procession from workshop to workshop, saying mantras and blessing the tools with purificatory fire. Behind him, Padmini walked proudly, in a long dress made out of scarlet cotton, a gift from Sister Felicity. The young Indian girl had persuaded the Scottish nun to join in the celebrations. When they spotted the cross round her neck, many of the workers asked her to come and bless their tools in the name of her god. 'Praise to you, oh God of the Universe, who gives our daily bread, for your children in Orya Bustee, Chola and Jai Prakash love and believe in you,' repeated Sister Felicity fervently in each workshop. 'And rejoice with them at this day of light in all the hardship of their lives.'

# 23

# 'Half a million hours of work and not a day lost'

The City of the Begums could not help but bless the Chairman of Union Carbide. No other industrial enterprise within Bhopal's ancient walls had been quite so concerned about its image; no other was quite so solicitous towards its staff. Each day brought new examples. In the plant, Muslim workers had the use of a place of prayer facing Mecca; Hindus had little altars dedicated to their principal gods. During the Hindu festival in honour of the goddess Durga, the management gave the workers a generator to light up her richly decorated statue. The material advantages were no less plentiful. A special fund enabled people to borrow money for weddings and festivals. The insurance and pension plans put the factory ahead of most Indian firms. A canteen, accessible to all, dispensed meals for a token price of two rupees.

In accordance with what they had been taught in Institute, however, it was the safety of their staff that was the prime concern of those in charge of the plant. Carbide equipped Hamidia Hospital with ultra-modern resuscitation equipment, which could treat several victims of gas poisoning simultaneously. The gift was greeted with public celebrations widely reported in the press. In addition a hospital infirmary with all kinds of respiratory equipment, a radiology unit and a laboratory, was built at the very entrance to the site. 'We were convinced all these precautions were unnecessary,' Kamal Pareek said afterwards, 'but they were part of the safety culture with which we had been inculcated.' Yet this same culture accommodated some surprising deficiencies. The medical staff engaged by Carbide did not have the benefit of any specific training in the pathology of gas-related accidents, especially those caused by methyl isocyanate.

⁂

Relaying the training received in Institute to over a thousand men who were oblivious to the extreme dangers they faced, was the task of the young assistant manager for safety. 'Making people appreciate the danger was virtually impossible,' Pareek recounted. 'It's in the nature of a chemical plant for the danger to be invisible. How can you instil fear into people without showing them the danger?' Meetings to inform people, emergency exercises, poster campaigns, safety demonstrations in which families took part, slogan competitions . . . Pareek and his boss were constantly devising new ways of awakening everyone's survival instinct. Soon, Warren Woomer was able to send a victory report to his superiors in America: 'We are pleased to announce that half a million hours have been worked without losing a single day.'

Safety, Pareek knew, also depended upon a certain number of specific devices, such as the alarm system with which the plant was equipped. At the slightest intimation of fire or the smallest emission of toxic gas, the duty supervisor in the control room had orders to set off a general alarm siren. At

154

the same time loudspeakers would inform personnel, first in English, then in Hindi, of the precise nature of the gas, the exact location of the leak and the direction in which the wind was blowing. This last piece of information was supplied by a windsock at the top of a mast outside the MIC unit. In case of a major leak, staff would receive an order to evacuate the site. Withdrawal from the factory would be carried out without panic, just like in the practices Pareek regularly organized.

All the same, this alarm system was only directed at the crews working on the factory site. None of the loudspeakers pointed outwards, that is to say in the direction of the bustees where thousands of potential victims were packed together. 'From the moment I got there, the proximity of all those people was one of my major worries,' Warren Woomer would recall. 'Every evening I would have our guards move away those setting up camp right on our perimeter wall. Sometimes some of them would even get under the fencing and we would have all the difficulty in the world getting them out. The plant had such magnetic appeal! So many people wanted to get a job there! That's what drew them nearer and nearer.' One day Woomer decided to intervene personally with the municipal government to get them to force people to 'move as far away as possible' from his installations. His efforts failed. None of the authorities appeared disposed to launch another eviction operation against the Kali Grounds' squatters. Woomer proposed drawing up a plan to evacuate people in case of a major incident. The very idea of such a plan provoked an immediate hostility from the very highest levels of the Madhya Pradesh government. Wasn't there a danger of throwing people into a panic, of driving some of them away? – a danger which Arjun Singh, the state's Chief Minister, did not, at any price, wish to run. The elections were approaching and he needed

155

every possible vote, no matter where it came from. The portly Omar Pasha, his electoral agent in the three bustees, was already campaigning on his behalf. Astute politician that he was, he had anticipated everything to ensure his re-election. Not only would he prevent the expulsion of his electors, but he would win their votes by giving them soon the best present they could ever dream of.

❧

The scene that engineer Kamal Pareek imagined one day was like a clip from a horror movie. The metal in one of the pipelines had cracked, allowing a flood of methyl isocyanate to escape. Because the accident was not the kind of leak the safety equipment could contain, the ensuing tragedy was unstoppable. A deadly cloud of MIC was going to spread through the factory, then into the atmosphere. The idea for this disastrous scenario came to Pareek as he watched a train packed with passengers come to a halt on the railway line which ran between the factory and the bustees. Would it be possible for a cloud of MIC driven by the wind to hit those hundreds of poor wretches trapped in their railway carriages? The engineer wanted to know. He went to Nagpur, former capital of the Central Provinces, and presented himself at India's national meteorological headquarters. In its archives were kept records of meteorological studies carried out in India's principal cities for the last quarter of a century: temperatures, hygrometric and barometric pressure, air density, wind intensity and direction and so on. All this information was recorded on voluminous rolls of paper. After a week spent compiling data, the engineer was able to extract from this ocean a mass of information about the meteorological conditions peculiar to Bhopal. For example, in 75 per cent of cases, the winds blew from

156

north to east at a speed of between six and twenty miles an hour. The average temperature in December was 15° celsius by day but only 7° celsius at night.

Pareek packed this paperwork in a cardboard box and dispatched it swiftly to the safety department at Union Carbide, South Charleston, to have it simulated on the computer. Taking into account the meteorological conditions prevalent in Bhopal, the technicians in the U.S. should be able to tell whether or not the toxic cloud of his scenario was likely to hit the train that had stopped next to the bustees. The reply came back three days later in the guise of a short telex. *It is not possible, even under the worst conditions, that the toxic cloud will hit the railway line. It will pass over it.*

'It will pass over it . . .' the engineer repeated several times, catching his breath. A vision of horror passed before his eyes. 'My God,' he thought, 'so the cloud would hit the bustees.'

The vigorous games of tennis Warren Woomer played every morning before going to his office reflected his high morale. The 'beautiful plant's' top man had every reason to be satisfied. After a mediocre first year, the production and sales of Sevin had taken off, to reach 2704 tons in 1981: half the factory's capacity but 30 per cent more than Eduardo Muñoz's most optimistic predictions prior to his departure. Despite this success, however, the 'beautiful plant' had grave problems. The most serious arose from the alpha naphtol production unit. The installation designed by Indian engineers had never, despite several modifications, been able to supply a product that was pure enough. They had therefore to resign themselves to importing alpha naphtol directly from Institute in the United States. In the end this fiasco would cost Carbide 8 million dollars, 40 per cent of the original budget for the entire plant.

To compound this misfortune, in 1978 a fire had devastated part of the alpha naphtol unit. The gigantic column of black smoke that hid the sun before raining down foul-smelling particles on roofs and terraces, had been Carbide's first signature in the sky over Bhopal. Seeing this incredible spectacle from his house, a young journalist by the name of Rajkumar Keswani rushed to the scene of the disaster, only to find that the area had already been cordoned off by hundreds of policemen. No one was allowed near.

Nonetheless, four years after this accident, Carbide's star continued to shine in the firmament over the City of the Begums. On Shamla Hill, the scientists in the research centre had just discovered a new molecule that was even more effective against predators attacking rice and cotton in the region. The guest house's panoramic restaurant overlooking the town had become the favourite meeting place of the political establishment and local society. Those who went there would never forget the extravagant spectacles that formed the after-dinner entertainment, like the water ballet in the swimming pool that the wife of the managing director of Carbide's Indian subsidiary, herself an accomplished dancer and swimmer, had arranged. The initiated knew that this luxurious residence was also used for top secret meetings. Carbide had placed a suite at the permanent disposal of Arjun Singh, the Chief Minister of Madhya Pradesh. In Bhopal, as elsewhere, money and power made comfortable bedfellows.

# 24

## *Everlasting roots in the black earth of the Kali Grounds*

The word had gone round like a trail of gunpowder from hut to shed to stall to workshop. The residents of the three bustees were to gather on the tea house esplanade for a communication of the utmost importance.

'This is it. Carbide's taking us all on!' clarioned Ganga Ram, who had never got over being rejected because of his mutilated hands.

'In your dreams, you poor fool!' Iqbal, ever the pessimist, called out to him. 'It's to inform us we're going to be evicted. And this time, it'll be for good!'

The arrival of Dalima on her crutches interrupted the exchange. With a yellow marigold in her hair, glass bangles jangling about her wrists, the young cripple had a triumphant air about her.

'It's to tell us they're going to install a drinking water supply with taps!' she announced.

'Why, it's obvious,' said old Prema Bai, 'they need us for the elections.'

In India, like anywhere else, it was the womenfolk who were most perspicacious.

That was when a voice from a loudspeaker rent the sky.

'People of Orya Bustee, Jai Prakash and Chola, hurry up!'
it commanded.

The residents of the bustees burst from the alleyways like tributaries joining a river flowing to the sea. Sister Felicity, who was in the process of giving several children polio vaccinations, paused.

'It's like being at home in Scotland when a storm breaks,' she told Padmini. 'All the sheep start running towards the voice that's calling them.'

Padmini made an effort to imagine the scene. She had never seen sheep. At that point Rahul, the legless cripple, appeared.

'Padmini, run to the factory and tell your father and the others. Ask him to round everyone up.' Suddenly assuming the mysterious air of one who knew more, he whispered: 'I think our state's precious Chief Minister has a surprise for us.'

The young Indian girl set off for the factory at a run. Everywhere the sweatshop slaves were abandoning their tools and their machines to make for the grand gathering. As they arrived Belram Mukkadam directed them to sit down with his stick. Soon the entire parade ground was covered by a human sea.

A truck appeared. It was loaded with posters that Mukkadam immediately had hung all around the tea house. On most of them, people recognized the balding forehead, fleshy lips and thick glasses of the Chief Minister of Madhya Pradesh. Other posters depicted an open hand. In the same way that Shiva had a trident as his emblem, Vishnu his wheel and the religion of Islam a crescent, the Congress Party, of which Arjun Singh was one of the leading lights, had chosen as its symbol the wide open palm of a hand. The truck was also carrying a collection of small fliers, which Rahul, Ganga Ram and others busied themselves distributing.

*WE LOVE YOU, ARJUN!* they said, *ARJUN, YOU ARE OUR SAVIOUR! ARJUN, BHOPAL NEEDS YOU!* Some of the bills even went so far as to proclaim: *ARJUN, INDIA WANTS YOU!*

Detained in New Delhi with Indira Gandhi, the organizer of this incredible show had entrusted his official representative in the Kali Grounds' bustees to see that the display served his electoral interests. The fact that the guest of honour was missing made the spectacle all the more quaint, for the proceedings began with the solemn arrival of an empty armchair. Carried by two servants in *dhotis*, the august seat came directly from the drawing room of Omar Pasha's residence. Encrusted with mother of pearl and ivory, it looked more like a throne. A few minutes later, a gleaming Ambassador brought the Chief Minister's representative. In honour of the occasion, Omar Pasha was wearing that most legendary crown in India's history, the white cap of those who had fought for independence. Thirty-eight years after the death of Mahatma Gandhi, the godfather of the bustees knew that that white cap was still a magical rallying sign.

Behind the old man walked, at a respectful distance, his son Ashoka, a tall fellow with a shaven head, whom the inhabitants of the bustees had learned to fear and respect. Manager of the clandestine drink trade controlled by his father, today he carried neither alcohol nor hashish, but a small ebony chest sealed with a copper lock. Inside this casket was a treasure, possibly the most invaluable treasure the occupants of Orya Bustee, Chola and Jai Prakash could ever hope for.

Omar Pasha sat down on his throne, in front of which Mukkadam had placed a table covered with a cloth, with a bouquet of flowers and incense sticks on it. Because of the brightness of the sun, the godfather's eyes were hidden

161

behind dark glasses, but people could tell what he was thinking by the way he wrinkled his eyebrows. Mukkadam called for a microphone, which the visitor seized between podgy fingers dripping with gold and ruby rings.

'My friends!' he exclaimed in a strong voice that forty years of cigar smoking had not managed to roughen, 'I have come to deliver to you, on behalf of our revered Chief Minster Arjun Singh . . .'

At this name, Pasha paused, sending a tremble through the assembly bristling with posters. Someone shouted: 'Arjun Singh, *Ki Jai*!', but the cry was not taken up. The crowd was impatient to hear the rest of the speech.

'At the request of our Chief Minister,' the godfather went on, 'I have come to deliver to you your *patta!*[1]'

The echo of this unbelievable, supernatural, unhoped for word, hovered in the overheated air for interminably long seconds. Surveying the stunned crowd, Sister Felicity could not help thinking of a sentence by the Catholic writer Léon Bloy: 'You don't enter paradise tomorrow, or in ten years' time. You enter it today when you are poor and crucified.'

Since the dawn of India's history that mythical word, 'patta' had haunted the dreams of millions of people, whose unfavourable karma had deprived them of the fundamental right to own a roof over their heads; it had fired the hopes of all those who, in order to survive, had had no alternative but to set up their hovel wherever they could. The people who had ended up in the Kali Grounds were among those poor unfortunates. Those people, whom Indira Gandhi's son had tried forcibly to drive away, those people whom the managing director of an American plant dreaded seeing encamped against its walls, had for years been clinging desperately to the pitiful patch of dust that Belram Mukkadam

---

[1] Property title deeds.

had one day traced out for them with his stick. And there, suddenly, was the godfather bringing them official property deeds issued by the government of Madhya Pradesh.

It was too good to be true. Never mind the fact that this deed would have to be renewed in thirty years' time; never mind the fact that they would have to pay 34 rupees a year in tax; never mind the fact that it was officially forbidden to pawn or sell it. A frenzied cheer went up from the crowd, which rose to its feet in a single movement. People chanted the names of Arjun Singh, Omar Pasha and Indira Gandhi. They danced, they laughed and congratulated one another. Caught up in a surge, Padmini suddenly found herself raised above the surrounding heads like a figurehead, the fragile emblem of a people throwing off its chains and achieving the beginnings of dignity. As far as these uneducated men, women and children were concerned, the pieces of paper Omar Pasha pulled from his chest were a gift from the gods. These deeds would remove their fears for good and allow them to plant their roots forever in the welcoming ground over which fluttered the flag with the blue and white logo.

Every time Omar Pasha invited a beneficiary to come and collect the document inscribed with his name and the designation of his plot, a bearded character sitting at the back wagged his head and rubbed at his enormous eyebrows. For the Sikh, Pulpul Singh, the neighbourhood usurer, this was a fortune on a plate, the opportunity to increase his wealth, even if it meant breaking the law in the process. Pulpul Singh could already see each sheet of paper that came out of the godfather's chest winging its way into his own safe. The day would come when these poor people needed to borrow money from him and what better guarantee could he ask for than the deposit of those magical deeds, which he would always find a way of selling on at a profit?

## PART TWO

# A NIGHT BLESSED
# BY THE STARS

# 25

## *A gas that makes you laugh before it kills you*

With his thick moustache, bushy eyebrows and round cheeks, the 32-year-old Muslim Mohammed Ashraf was the mirror image of the Indian cinema idol Shashi Kapoor. The resemblance had made him the most popular worker in the plant. In charge of a shift in the phosgene unit on that 23 December 1981, Ashraf had to carry out a routine maintenance operation. It was a matter of replacing a defective flange between two pieces of pipework.

'No need to put your kit on today,' he announced to his colleague Harish Khan, indicating the heavy rubber coat hanging on a hook in the cloakroom. 'The factory isn't running. There's no likelihood of a leak.'

'Gases can go walkabout even when everything's stopped,' retorted Khan sharply. 'Better be on the safe side. A few drops of that blasted phosgene on your pullover can hurt you. It's not like the *bangla*[1] from Mukkadam's tea house!'

The two men burst out laughing.

'I'm willing to bet Mukkadam's gut-rot is even more

---

[1] Alcohol distilled from fermented animal guts.

167

dangerous than this bloody phosgene,' concluded Ashraf, in a hurry to put on his mask.

No one had ever had cause to reproach the Muslim operator for any breach of safety procedures. Ashraf was one of the most reliable technicians in the company, even if he did leave his workstation five times a day to go out into the courtyard and pray on his little mat facing Mecca, and even if he did come staggering to work in the morning because he had spent all night on the banks of Upper Lake fishing. The son of a small trader in the bazaar, he owed everything to Carbide, not least his marriage to the daughter of a cloth merchant from Kanpur, who was honoured to have an employee of the prestigious multinational for his son-in-law, even if he was only a low level employee. A graduate in economics, Sajda Bano was a beautiful young woman. She had given him two sons, Arshad and Soeb, in whom he could already see two prospective 'Carbiders'.

It took only a few minutes to dismantle the joint. Just as he was fitting the new component, however, Ashraf saw through his mask a small quantity of liquid phosgene spurt from the upper side of the piping. A few drops landed on his sweater. Aware of the danger, he rushed into a shower cabin to rinse his clothing. It was then that he made a fatal mistake. Instead of waiting for the powerful jet of water to complete the decontamination process, he took off his mask. The heat of his chest immediately caused the few drops of phosgene still nestling in the wool of his sweater to vapourize. Apart from a slight irritation of the eyes and throat, which rapidly disappeared, Ashraf felt no discomfort at the time. He did not know that phosgene has a Machiavellian way of killing its victims. First it gives them a sense of euphoria. 'I'd never seen my husband so voluble,' Sajda Bano would recall. 'He seemed to have forgotten the accident. He took us out in the car to visit a small country house he wanted

168

to buy beside the Narmada River. He was as cheerful as he was during the first days of our engagement.' Then, all of a sudden, he collapsed, with his lungs full of a fierce flood of secretions. He started to vomit a gush of transparent fluid mixed with blood. Panic-stricken, Sajda called the factory, who had him taken by ambulance to the intensive care unit Carbide had donated to Hamidia Hospital. He was placed on an artificial respirator. His agony went on. He threw up more and more secretions, up to four and a half pints an hour. Soon he did not even have the strength to expectorate.

Sajda had to push aside some of her in-laws' family to get to her husband's bedside. 'He was as white as a sheet,' she remembered, 'but when he sensed my presence, he opened his eyes and tore off his oxygen mask. "I'd like to say goodbye to the children. Go and fetch them!" he murmured.'

When the young woman came back with the two boys, the dying man took the youngest in his arms. 'Son, how do you fancy a fishing trip?' he asked, forcing a smile. The effort set off a violent bout of coughing. Then came a succession of rattles and a last sigh. It was all over. Bhopal's 'beautiful plant' had claimed its first victim. It was Christmas Day. For the young woman who had come from a far distant province to marry a Carbide man, three months and thirteen days of mourning were about to begin.

✳

The entire factory grieved for its martyr. One of those most affected by the accident was its Managing Director. 'We had nothing to reproach ourselves for,' Warren Woomer would remember. 'Mohammed Ashraf had been properly trained for the dangers of his profession. By neglecting to put on his rubber coat and taking off his mask too soon, he committed the worst of offences. It was the first time in my life

169

as an engineer that I'd lost one of my men. I'd had people injured but never a death. It was the kind of situation where you had to know exactly what had gone on because it must never be allowed to happen again. No matter what the circumstances of the accident.'

�֍

Two employees took it upon themselves to provide a response to the Managing Director's question. A Hindu, Shankar Malviya, and a Muslim, Bashir Ullah, led the firm's main trade union. Both came from very poor families in the Bhopal bustees. Their energy and readiness to intervene on behalf of their comrades had made them immensely popular. In a strongly worded letter, they formally accused the management of responsibility for their comrade's death. There was in fact a safety regulation prohibiting the storage of phosgene while the unit manufacturing it was not in production – as was the case at the time of Ashraf's accident. There should not have been any gas in the pipes whatsoever. Yet, despite this regulation, a quantity of phosgene had been left in the tanks. No one in authority had warned the unfortunate operator, a fact which rendered the company responsible. According to the two trade union leaders, the accident pointed irrefutably to a decline in safety standards at Carbide. They were, therefore, going to ask the government of Madhya Pradesh to place the factory immediately in the category of those manufacturing high risk products, which would subject them to much stricter safety requirements.

'For the first time people became aware of something that all our safety campaigns had been unable to make them appreciate: that the substances they were handling were deadly,' Kamal Pareek said. 'But this time the danger had a face to it.'

170

On 10 February 1982, a little over a month after the death of Mohammed Ashraf, another accident occurred involving the poisoning of twenty-five workmen who were rushed to hospital. Fortunately there were no deaths to mourn. Gas had leaked from a phosgene pump. The fact that none of the victims had been ordered to wear protective masks while moving about in a sensitive area further outraged the two trade union leaders. The management defended itself by stating that leaks resulting from mechanical failure never exceeded the toxicity level, above which such incidents were likely to be fatal. Malviya and Ullah tried in vain to find out how and according to what criteria this 'level', which did not feature in any of the company's manuals or official documents, had been determined. 'It was one of many mysteries surrounding Carbide's procedures in Bhopal,' they stated.

Nor was their fury likely to abate. On 5 October of the same year, a further accident struck the factory in the middle of the night. This time it occurred in the unit producing methyl isocyanate. As an operator was opening a valve in an MIC pipeline, the joint linking it to several other pipes unexpectedly broke, releasing a huge cloud of toxic vapours. Before evacuating the unit, the operator set off the alarm siren. A few seconds later, in accordance with the procedures set down by Kamal Pareek, the voice of the supervisor in the control room ordered a full evacuation of the plant. The position of the sock on top of the mast indicated a moderate wind blowing north-north-east. All those inside the factory made off as fast as their legs would carry them in the opposite direction, towards the Kali Grounds' bustees.

❧

171

The Mangala Express regularly wrested the residents of these bustees from their slumbers. They all dreaded the din it made as it went by. The only one not to suffer from it was Prema Bai, the elderly midwife, and that was because she was deaf. Her neighbours maintained that the noise of a herd of elephants trampling the hovels in her alleyway would not wake her. Yet it was she who gave the alert that night.

'Get up! Everybody up! There is pandemonium at Carbide!' she shouted, running from hut to hut, her white widow's sari flying out behind her.

Prema Bai was the first to have heard the distant howl of the siren. Awakened by her cries, the neighbours got up one by one, grumbling. They were angry at being woken for a second time. Everyone strained their ears in the direction of the muffled howling coming from the plant.

'Perhaps someone's set fire to it somewhere,' grinned Ganga Ram, who harboured a deadly hatred of Carbide.

'Calm down, friends!' intervened Belram Mukkadam. 'We hear that siren nearly every day. It's not sounding for us but for the guys inside the factory.'

'It may even have set itself off,' ventured Ahmed Bassi.

'All the same it's going,' interrupted Salar, the bicycle repairman, 'we should find out more.'

'You're right, Salar,' agreed Nilamber, fiddling nervously with his goatee.

A voice then rose up from the ground. Rahul had just arrived on his board. He had taken the time to arrange his bun and put on his necklaces.

'Look here, my friends, why does that siren frighten you?' he asked. 'We hear it nearly every day!'

'Yes, but, tonight, it's sounding without stopping,' interjected Sheela, Padmini's mother, visibly disturbed.

The crowd was growing by the minute. People who had been rudely awoken were arriving from Chola and Jai

Prakash, tousle-haired and barefoot. Old Prema Bai's cry had spread from alley to alley.

Ratna Nadar, Padmini's father, bent down to the legless cripple.

'Do you know what they make in that factory of Carbide's?' he enquired.

Rahul appeared surprised by the question.

'We should be asking you. You've been working there every day for two years.'

The little man appeared to think, then shrugged his shoulders in a gesture of helplessness.

'No, I've no idea. No one's ever told us.'

Rahul moved his plank forward into the middle of the assembly which formed a circle round him. His reputation as the bustees' best informed man commanded attention.

'Well, I'm going to be the one to tell you what Carbide is making on the Kali Grounds,' he declared. 'I've asked some big shots and I can assure you there's nothing to be afraid of. Carbide makes medicines for sick plants. Small white granules to get rid of the insects that attack them and steal the harvest off the poor bastards who planted them. And little white granules aren't dangerous to anybody. Except the blasted little creatures in the plants.'

Ratna Nadar could still see the hordes of black aphids that had devoured his field in Mudilapa.

'You mean to say that all those pipes, all that machinery, all those sacks of powder that go off on trucks, are just to kill those bloody little . . .'

His throat was constricted with emotion.

'You've got it, brother,' confirmed Rahul. Pointing his right hand, with its fingers covered with rings, at the illuminated factory, he assumed a solemn tone: 'You can go back to bed, friends. That siren isn't for us!'

Scarcely had the legless cripple finished speaking than five

173

men surged out of the darkness beside the railway track. Haggard, ghastly, exhausted, with their eyes starting out of their heads, they looked like spectres in a horror film. One of them was dragging an unconscious comrade. Other escapees came up behind that first group.

'Get out of here! There's been an accident,' gasped one of them who had stopped to recover his breath. 'The plant's full of gas. If the wind starts to blow in this direction, you're all for it.'

Belram Mukkadam raised his stick above the heads about him. He had tied his *gamcha*[2] to it and was waving it about like a flag to rally people.

'We're moving out!' he cried. 'Follow me! Quickly!'

The semblance of a procession formed behind him. No one panicked because, for all the howling of the siren, it was still difficult to believe in the danger. Before leaving, old Prema Bai lit incense before the image of God Jagannath on the small altar at the end of the alleyway. It was then that a pot-bellied individual with a shaggy beard and a scarlet turban appeared. With the help of his two sons, the usurer Pulpul Singh was carrying his most precious possession. He would never have left home without the safe to which he alone knew the combination.

✳

The fact that the alarm siren kept going off did nothing to shatter the confidence of those in charge of the factory. The local government authorities wrote to the two trade union leaders to assure them that the safety of the Carbide workers

---

[2] A piece of cotton used as a towel or as a scarf against the cold.

'would be subject to close investigation at the opportune moment'.

With the exception of the unfortunate Ashraf, the accidents had claimed no victims either inside or outside the factory. At Carbide they were seen, therefore, as the teething problems experienced by any new plant. The two trade unionists did not share this opinion. They had 6000 notices printed, which their members rushed to post on the walls of the factory and all round the town. *BEWARE! BEWARE! BEWARE! ACCIDENTS! ACCIDENTS! ACCIDENTS!* they protested in enormous red letters, under which was written: 'The lives of thousands of workers and hundreds of thousands of residents of Bhopal are in danger because of the toxic gases produced by Carbide's chemical plant.' The notices listed all the accidents that had occurred, the repeated violations of labour laws, the stretching of the safety standards. In order to really mobilize public opinion, however, Malviya was counting on a much more effective weapon. Mahatma Gandhi had successfully used it to induce the British colonialists to agree to his demands. It consisted of offering one's life to one's enemy. Malviya announced that he was embarking on a hunger strike.

# 26

## 'You will all be reduced to dust'

So the frail little man with the dark skin had actually dared to do it. For a week he had lain stretched out on a piece of *khadi*[1] cloth outside the entrance to the factory. With the nape of his neck resting on a stone, a pitcher of water beside him, he was the embodiment of the Carbide workers' revolt against the working conditions that had led to the death of one of their comrades. Every morning at dawn, five workers took their place beside Malviya to fast with him for twenty-four hours. Before going to their work-stations, the other employees would gather round the strikers to show their solidarity. '*Har zor zulm key takkar mein sangharsh hamara nara hai!* We will fight against all forms of oppression!' hundreds of voices shouted in unison.

For the multinational that had built a large part of its reputation on the slogan 'safety first', these hunger strikes and accompanying demonstrations were unacceptable blackmail. The reaction was swift and drastic. All political and trade union meetings inside the factory were banned. D. S. Pandi, the dynamic head of personnel, had no reservations

---

[1] Course cotton material spun on a wheel.

about going out and setting fire to the tent which served as the union's headquarters. In the ensuing scuffle several people were injured, among them Pandi himself, with the result that the trade union leaders were promptly laid off. Without renouncing the fight, they kept up their action outside. Meetings and processions denouncing the death of Mohammed Ashraf and demanding better safety standards were held one after another throughout the city, seriously denting the company's unanimously respected image in the public eye. Curiously, neither Woomer, nor his Indian assistants appeared unduly alarmed at this fierce outbreak of discontent. After all, wasn't this kind of labour unrest to be expected in Indian firms, where workers had been known to lock their bosses in their offices for weeks? But at Carbide, the fact that an ordinary worker could lie down on the pavement and defy the world's third largest chemical giant felt like a crime of lese-majesty; a crime which cast an unjust slur upon the ideal of 'giving India's peasants a hand' dreamt of in New York; a crime which destroyed the myth that belonging to Carbide was the best possible sign of a prosperous karma; a crime which diminished the prestige of the blue and white uniform that a whole generation of young Indian graduates dreamed of wearing. 'I knew the factory wasn't perfect,' Warren Woomer would say, 'but we were constantly improving it. Until Ashraf's death we'd had an excellent safety record, unique in the company's history.' The American could see no reason why this situation should deteriorate. He had blind faith in his colleagues. After all it was he who had trained them in Carbide's celebrated safety culture. He knew that the 400 pages of notes they had compiled on their return from Institute were their bible. A man's death was a dreadful blow but it should not cast disgrace upon the whole system. Despite the budget cuts to some of the equipment at the time of construction, Woomer was convinced that he was commanding one of the safest ships in the modern industrial fleet. Amongst the factory management, no one was in any doubt: these demonstrations were just a campaign by agitators wanting higher salaries and shorter working hours.

❋

It was one of thousands of weekly newspapers that India published in its innumerable languages. Bhopal's *Rapat Weekly* came out in Hindi, and its modest circulation – 6000 copies – gave it very little impact in a mostly Muslim city where the predominant language was Urdu. The reliability of its investigative journalism and the independence of its comment had, nevertheless, earned the *Weekly* a fringe readership with a taste for scandal. Digging into the latter was the particular slant the founder and only editor of the *Rapat Weekly* had chosen.

The son and grandson of journalists, the 34-year-old Hindu Rajkumar Keswani belonged to a family originally from the province of Sindh, who had come to Bhopal after the Partition of India in 1947. At sixteen he had left college to contribute to a sports journal, then worked on the news-in-brief column of the *Bhopal Post*. For years this indefatigable investigator had reported on the minor and major events that occurred in the City of the Begums. After the failure of the *Post*, Keswani had sunk his savings into the creation of this small weekly to serve the true interests of its citizens. He was a man who was mad about poetry, botany and music, and he felt modern industry posed a very real threat to the safety of the city. The discovery of irregularities in the allocation of industrial licences drove him to look for collusion between Carbide and the local authorities. The mysterious fire in the alpha naphtol unit had already tickled his curiosity. The poisoning of Mohammed Ashraf clinched the matter. He embarked upon an enquiry that might have turned him into a saviour if only people had listened to him.

'As luck would have it, I knew Ashraf,' he explained. 'He lived just next door to the fire station where I'd set up my office. He often had comrades from work round to his house. Together, they would talk about the dangers of their profession. They spoke about toxic gases, deadly leaks and the likelihood of explosion. Some of them made no secret of their intention to resign. I'd thought the plant was producing an innocent white powder like the one I used to protect the roses on my terrace from green-fly, and I found what they said terrifying.'

No sooner had he carried his friend to his grave than the journalist rushed to see the deceased's colleagues. 'I wanted to know whether his death was an isolated incident or the result of some failure on the part of the factory.'

Keswani gathered enough witness statements to bring an accusation against Carbide. Bashir Ullah, one of the dismissed trade union leaders, even managed to smuggle the journalist inside the site at night. As he went through the various production units, he could smell phosgene's odour of freshly cut grass and methyl isocyanate's aroma of boiled cabbage.

Not having any scientific training, he next paid a visit to the Dean of the chemistry department at an important technical college and consulted all the specialist works in its library. The conclusions he came to made his blood run cold. 'Merely appreciating that methyl isocyanate and phosgene are two and a half times heavier than air, and have a tendency to move along at ground level in small clouds, was enough to make me realize at once that a large-scale gas leak would be disastrous,' he explained. 'After detailed examination of the safety systems in place in the plant, I knew that tragedy was only a matter of time.'

✻

An unexpected visit was to provide Rajkumar Keswani with the technical arguments he needed to drop his journalistic bombshell. In the month of May 1982, three American engineers from the technical centre for the chemical products and household plastics division in South Charleston, landed in Bhopal. Their task was to appraise the running of the plant and confirm that everything was functioning according to the standards laid down by Carbide. It was a routine procedure, which no one expected to produce any great revelations.

The hundred or so breaches of operational and safety regulations the investigators turned up might not, at first glance, seem excessive in a plant as vast and complex.

With the help of accomplices in the factory, however, Keswani managed to get hold of the text of the audit. He could not believe his eyes. The document described the surroundings of the site 'strewn with oily old drums, used piping, pools of used oil and chemical waste likely to cause fire'. It condemned the shoddy workmanship on certain connections, the warping of equipment, the corrosion of several circuits, the absence of automatic sprinklers in the MIC and phosgene production zones, the risk of explosion in the gas evacuation flares. It cited the poor positioning of certain devices likely to trap their operators in case of fire or toxic leakage. It criticized the lack of pressure gauges and the inadequate identification of innumerable pieces of equipment. It reported leaks of phosgene, MIC and chloroform, ruptures in pipework and sealed joints, the absence of any earthing wire on one of the three MIC tanks, the

180

impossibility of isolating many of the circuits because of the deterioration of their valves, the poor adjustment of devices where excessive pressure was in danger of allowing water into the circuits. It marvelled at the fact that the needle on the pressure gauge of a phosgene tank full of gas was stuck on zero. It expressed alarm at the poor state and inappropriate placement of safety equipment to be used in case of leakage or fire, and at the lack of periodic checks to ensure sophisticated instruments and alarm systems were functioning correctly.

All the same it was in the human domain that the report came up with the most startling revelations. It expressed concern at an alarming turnover of inadequately trained staff, unsatisfactory instruction methods and a lack of rigour in maintenance reports. Three lines in the fifty-one pages pointed to a particularly serious mistake: an engineer had cleaned out a section of pipework without blocking off the two ends of the pipe with discs designed to prevent the rinsing water from seeping into other parts of the installation. One day this piece of negligence would spark off a tragedy.

⁂

'KINDLY SPARE OUR CITY!' exclaimed Rajkumar Keswani in the title of his first article, published on 17 September 1982. Illustrating the risk the factory represented with numerous examples, the journalist appealed first to those running it. 'You are endangering our entire agglomeration, starting with the Orya Bustee, Chola and Jai Prakash districts nestling against the walls of your installations.' Then addressing his fellow citizens, Keswani urged them to wake up to the danger that Union Carbide represented to their lives. 'If one day disaster strikes,' he warned them, 'don't say that you did not know.'

Poor Keswani! Like Cassandra, he had been given the gift of predicting catastrophe, but not that of persuasion. His first article passed almost unnoticed. Carbide was too well ensconced on its pedestal for a few slanderous words in a sensationalist newspaper to topple it.

Undaunted, the journalist returned to the fray two weeks later. 'BHOPAL: WE ARE SITTING ON A VOLCANO' announced the *Rapat Weekly* of 30 September 1982 in block capitals right across the front page. 'The day is not far off when Bhopal will be a dead city, when only scattered stones and debris will bear witness to its tragic end,' the author prophesied. The article's disclosures should have sent the entire city rushing to the Kali Grounds to demand the plant's immediate closure. They did not. Sadly, the *Rapat Weekly* was a voice crying in the wilderness.

The following week, a third article entitled, 'IF YOU REFUSE TO UNDERSTAND, YOU WILL BE REDUCED TO DUST', described in detail the leak which, four days earlier, had led to the evacuation of the factory in the middle of the night and the general scramble on the part of the residents of Orya Bustee and its adjacent neighbourhoods.

In the end so much indifference and blindness disheartened the journalist. If the Bhopalis preferred to believe the lies spread by Carbide's propaganda, he would leave them to their lot. He scuttled his newspaper, packed his music collection into two suitcases and bought a train ticket for Indore where a big daily offered him a golden opportunity. Before he left Bhopal, however, he wanted to respond to the Employment Minister for the state of Madhya Pradesh, who had just announced on the parliamentary rostrum: 'There is no cause for concern about the presence of the Carbide factory because the phosgene it produces is not a toxic gas.' In two long letters, Keswani summarized the findings of his personal investigations. He addressed the first to the state's highest authority, Chief Minister Arjun Singh, whose links with the directors of Carbide were common knowledge. The second he sent to the President of the Supreme Court along with a petition requesting the closure of the factory. Neither took the trouble to reply to him.

# 27

## Ali Baba's treasure
## for the heroes of the Kali Grounds

'Everyone to the tea house! Ganga has a surprise for us!'

Rahul sped along like lightning on his plank on wheels, bearing the news from alleyway to alleyway. Orya Bustee, Chola and Jai Prakash at once emptied themselves of their occupants. Their vitality and their incredible ability to mobilize themselves were the hallmarks of these disinherited people. With each of her weekly visits, Sister Felicity became more and more convinced that the poor she came to help were stronger than any misfortune.

The man who was promising them 'a big surprise' today was one of the most respected characters in the three bustees. With the passage of the years, Ganga Ram had become, like Belram Mukkadam and the godfather Omar Pasha, one of the Kali Grounds' influential figures. His rejection by Carbide's *tharagar* a few years previously had not dented his staying power. That same year, a few days before Diwali, when all Hindus repaint their houses, Ganga had turned himself into a house painter. In order to buy himself a ladder, a bucket and some brushes, he had paid a visit to another leprosy survivor, whom he had helped during his tenure at

Hamidia Hospital. Welcomed as if he were the god Rama in person, Ganga had been able to borrow the money he needed. Two years later his business had six employees. Success had not gone to his head, however. Ganga Ram had not left the neighbourhood where once four lines drawn with a stick in the dust had provided him and his small family with shelter. The whole community valued his wife Dalima, this bright young woman with her green eyes and her tattooed hands, who got about on her crutches without complaining and always with a smile. Modest in the extreme, she never lifted the bottom of her sari to reveal the horrible scars on her legs and the fractured bones that stood out beneath her skin. Frightened by the gangrene that was spreading through her legs, the surgeon at Hamidia Hospital had wanted to amputate. The young woman's reaction had been so passionate that it had roused the entire hospital. 'I'd rather die than lose my legs!', she had informed the surgeon. So he had tried giving her a metal pin and a bone graft, and Dalima had managed to keep her legs, but they were lifeless. The poor woman would have to haul herself round on crutches for the rest of her life or allow herself to be carried by the former leper to whose destiny she had been lucky enough to join her own.

Ganga Ram organized the surprise at the tea house as if it was a festival. He had exchanged his sandals and old blue painting shirt for gondola-shaped mules and a magnificent *kurta* of embroidered white cotton. Giving free rein to his comic talents, he had dug out a top hat, which made him look like the ringmaster of a circus. Six musicians wearing red cardboard shakos on their heads, and yellow waistcoats with Brandenburgs over white trousers, stood round him. Two of them held drumsticks between stumps once eaten away by leprosy, two more held cymbals, and the others, dented trumpets. Santosh, one of the trumpeters, a jolly little man with a face pitted from small pox, was Dalima's father. He had arrived from Orissa, where that year a drought even more severe than the ones from which Padmini and her family had suffered, was raging.

Just like on the day when the property deeds were handed out, Ganga Ram, Mukkadam, Salar the bicycle repairman and all the other members of the usual team directed arrivals to sit down in a semicircle round the tea house. When there was no more room, Ganga greeted the crowd and signalled to the musicians to break into the first piece. A necessary part of any Indian public gathering, much to everyone's delight, a raucous din immediately enveloped the assembly. After a few minutes, Ganga raised his top hat. The music stopped.

'My friends!' he exclaimed, 'I've gathered you together to share in an event so happy I couldn't keep it to myself. Now you're all here, I'm going to fetch the "surprise" I have for you.'

He signalled to the musicians to clear a way for him. A few moments later, the little procession was back, to a cacophony of trumpets, a roll of drums and the crash of cymbals. Behind the musicians walked the former leper with all the majesty of a Mogul emperor. He was carrying his wife Dalima who was draped in a blue muslin sari embroidered with gold patterns. With her tattooed wrists and the pendant earrings shining in her ears, the young woman was smiling and greeting people with all the grace of a princess. When the procession arrived outside the tea house, Ganga and the musicians turned to face the crowd. The din of the trumpets and cymbals increased by another few decibels.

With a nod of his head, Ganga stopped the music. Next, throwing out his chest like a fairground athlete, he held his wife out at arm's length as if presenting her as a gift to the crowd. Then with a face flushed with pride, he allowed Dalima to slip gently down to the ground. As soon as her feet touched the earth, she straightened up with a thrust of her loins and, cautiously, began to walk. Bewildered and completely at a loss, the people present could not believe their eyes. The woman whose silent torture they had all witnessed for so many years was there, before them, fragile and tottering, but on her feet. People stood up to get a closer look at the woman who had been so miraculously healed. Her husband had thought of everything; garlands of sweet-smelling yellow marigolds appeared. Padmini and Dalima's son, Dilip, strung more flowers round her neck. Soon the young woman disappeared beneath a pile of garlands engulfing her from her shoulders to the top of her head. Ganga was crying like a baby. He brandished his top hat to speak to the assembly again.

'Brothers and sisters, the celebrations are only just beginning,' he cried in a voice choked with emotion. 'I have a second surprise for you.'

This time, it was young Dilip who went off with the band to fetch Ganga Ram's latest 'surprise'. Dilip no longer 'did' the trains. He was now a sturdy young man of eighteen who worked as a painter with his stepfather. He was known to have only one passion: kite-flying. Every day, his aircraft made out of paper and rags carried aloft all the fantasies for freedom and escape of a trapped people.

What the former leper would give his companions that day was a rather different means of escape. Preceded by the six musicians bellowing out a triumphal hymn, Dilip returned, carrying on his head a rectangular shape concealed beneath a red silk cloth. Dalima followed her son's progress with the anxiety of an accomplice. Ganga ordered the young man to put the object down on a table that Mukkadam had prepared for the purpose. His mischievous smile betrayed how much he was enjoying his position. Again he silenced the music and took up his top hat.

'My friends! Can any one of you tell me what's under this cloth?' he asked.

'A chest to keep clothes in,' cried Sheela Nadar, Padmini's mother.

Poor Sheela! Like most of the other bustee families, hers had no furniture. A rusty tin trunk, often overrun with cockroaches, was the only place she had in which to keep her wedding sari and her family's few clothes.

A little girl went up and pressed her ear to the 'surprise'. 'I bet you've got a bear shut in a cage under your cloth.'

Ganga burst out laughing. The child's guess was not that unlikely. In Orya Bustee, as in all the other neighbourhoods, rich and poor alike, animal exhibitors and other showmen were not unusual. Trainers of monkeys, goats, mongooses, rats, parrots and scorpions; viper and cobra charmers . . . at any moment, a handbell, a gong, a whistle or a voice might announce the passing of some spectacle or other. The palm of honour went to bear trainers, especially as far as the youngsters were concerned. Giving the children of Orya Bustee a bear would certainly have been a marvellous idea. But Ganga Ram had had an even better one. With all the care of a conjuror about to produce doves, he placed his top hat on the mysterious object. Then, clapping his hands, he gave the band its signal. The drums and cymbals mingled with the trumpets in a deafening cacophony. As if for some ritual, Ganga then invited Dalima to walk three times round the table on which his 'surprise' was sitting. Proud and erect under her veil of blue silk bordered with a golden fringe, the young woman proceeded cautiously. Her steps were still unsteady but no one could take their eyes off her. They were hypnotized, for, at that instant, she was the embodiment of the determination of the poor to triumph over adversity.

As soon as Dalima had completed her three circuits, Ganga continued.

'And now, my friends, Dalima herself is going to unveil my second surprise,' he announced.

When the young woman tugged at the cloth, an 'oh!' of amazement burst from the throats of all those present. Nearly ten years after their country had sent a satellite into space, six years after they had set off an atomic bomb, tens of millions of Indians did not even know such a device existed. Enthroned on the tea house table sat the Kali Grounds' first ever television set.

Those in charge of the 'beautiful plant' sat down round the long teak conference table to examine the crushing report by the three auditors from South Charleston. Kamal Pareek, Assistant Manager for Safety, felt particularly concerned about some of the findings. He was not alone. 'The anomalies the report revealed might well have been part of the usual teething problems of a large plant,' he maintained later, 'but they were still serious.' That was also the opinion of the American Managing Director. Warren Woomer belonged to a breed of engineers for whom one single defective valve was a slight upon the ideal of discipline and morality that ruled his professional life. 'Not tightening a bolt properly is as serious an offence as letting a phosgene reactor get out of control,' he would tell his operators. In his quiet, slightly languid voice he enumerated the report's observations. Before seeking out the guilty and sanctioning them, all the anomalies had to be rectified. That could take weeks, possibly even months. A schedule for the necessary repairs and modifications to the plant would have to be sent to the technical centre in South Charleston and approved by its engineers.

It would fall to a new captain to bring the Bhopal factory back up to scratch, however. In its desire to proceed with the complete Indianization of all foreign companies in their country, the New Delhi government had declined to renew Woomer's residence permit. His replacement, a 45-year-old Brahmin with the swarthy skin of a southerner, and an impressive academic and professional record, was already sitting opposite him. The Chairman of Carbide and his board of directors had unreservedly approved the appointment of this exceptionally gifted individual. Yet, in the space of two years, Jagannathan Mukund was to steer the Bhopal factory to disaster.

※

Once more the people of the bustees demonstrated their resourcefulness. In less than an hour Ganga Ram's television set was broadcasting its first pictures. In the absence of any electricity in the neighbourhood, Ganga Ram's friends had run a cable to the line that supplied the factory. Salar had rigged up an aerial with a wheel mounted on a bicycle fork. The pirate apparatus had a very superior look to it, like a satellite listening station.

Suddenly a picture lit up the screen. Hundreds of eyes nearly jumped from their heads as they watched a Hindi newsreader announce the programme for Doordarshan, the national television network. At a single stroke that picture banished all the greyness, mud, stench, flies, mosquitoes, cockroaches, rats, hunger, unemployment, sickness and death. And the fear too, that the great factory with its strings of light bulbs illuminating the night, had begun to inspire.

Every evening the programme on Indian television's only

channel began with the latest episode in a serial. The epic of the Ramayana is to India what the *Golden Legend*, the *Song of Roland* and the Bible are to the West. Thanks to Ganga Ram, the occupants of the Kali Grounds were going to watch the thousand dramas and enchantments of their popular legend unfold before them. Every evening for an hour, they would live out the marvellous love story of Prince Rama and his divine Sita. They would laugh, cry, suffer and rejoice along with them. Many of them knew whole passages from it by heart.

Padmini could remember how, when she was little, her mother used to sing to her the mythical adventures of the monkey general. Later, whenever storytellers passed through her village, her family would gather with the neighbours in the square, to listen to the fantastic stories which, ever since the dawn of time, had imbued everyday life with a sense of the sacred. No baby went to sleep without hearing its elder sister intone some episode from the great epic poem. Children's games were inspired by its clashes between good and evil, schoolbooks exalted the exploits of its heroes, marriage ceremonies cited Sita's fidelity as an example. Bless you, Ganga Ram, for thanks to you it was possible to dream once more. Seated before your magic lamp, the men and women of the Kali Grounds would be able to draw new strength to surmount the tribulations of their karma.

# 28

## *The sudden arrival of a cost-cutting gentleman*

Fourteen years, six months and seventeen days after an Indian mason had laid the first brick of the Bhopal Carbide factory on its concrete foundations, its last American captain left. 'That 6 December 1982 will always be one of the most nostalgic days of my life,' Warren Woomer would remember. The week prior to their departure, the Woomers were caught up in a whirlwind of receptions. Everyone wanted to bid farewell to the 'quiet American', who had known how to marry up the different cultures in his Indian workforce with the requirements of a highly technological industrial plant. It was true that the death of Mohammed Ashraf, the trade union unrest at the beginning of the year, and the worrying conclusions of the summer audit had revealed some cracks in the ship. But *Sahb*, as the Indian workers affectionately called him, left with his head held high. All the problems would be resolved, the bad workmanship would be rectified, the gaps filled. He was convinced that no serious accident would ever tarnish the reputation of the 'beautiful plant' in the heart of the subcontinent. It would continue to produce, in total safety, the precious white powder that was indispensable to

India's peasants. Woomer accepted the gifts engraved with his name with gratitude.

The American did know, however, that there were only two conditions upon which the factory could have a trouble-free future. The first was the favourable disposition of the Indian sky. Without generous monsoons to produce abundant harvests, the peasants would be unable to buy Sevin, in which case production would have to be slowed down and possibly even stopped for long periods. The financial consequences of such events would be grave. The other condition was compliance with the safety regulations. Woomer had talked at length to his successor about this. Throughout his long career dealing with some of the most toxic chemical substances, he had expounded a philosophy based on one essential principal: 'Only ever keep a strict minimum of dangerous materials on site.' With this profession of faith, the engineer was indirectly criticizing those who, against the advice of Eduardo Muñoz, had taken the risk of installing three enormous tanks capable of containing more than 120 tons of methyl isocyanate. 'I left with the hope that those tanks would never be filled,' he said, 'and that the small quantity of gas stored to meet the immediate needs of Sevin production would always be rigorously refrigerated as prescribed by the manual compiled by the MIC specialists.'

Like all lovers of culture, art and beauty, Warren Woomer and his wife Betty had succumbed to the magic of India. They promised themselves that they would return. The American was not aware of Rajkumar Keswani's articles. None of the Indians who worked for him had mentioned them. Looking back for one last time at his 'beautiful plant' through the rear window of the car taking him to the airport, Woomer wished it 'good luck'.

✻

The first sign that drought was once again afflicting the countryside of Madhya Pradesh and its bordering states was the sudden appearance of destitute families on the outskirts of the Bhopal bustees. A massive influx of 'untouchables', the outcastes whom Gandhi had baptized 'Harijans – Children of God' was the first hint that not a single grain of rice or ear of corn would be harvested from the fields that year.

Belram Mukkadam, the members of the committee for mutual aid and all the other residents set about making the newcomers welcome: one person would bring a blanket, someone else an item of clothing, a candle, rice, oil, sugar, a bottle of paraffin, or matches. Ganga Ram, Dalima and her son Dilip, Padmini and her parents, the old midwife Prema Bai, the godfather Omar Pasha with his two wives and his sons, the sorcerer Nilamber, the shoemaker Iqbal, the tailor Bassi, and the legless cripple Rahul were, as always, the first to show their solidarity. Even the sons of the usurer Pulpul Singh brought food for the refugees. Seeing all these people sharing what they had, Sister Felicity, who had come running with her First Aid kit, thought: 'A country capable of so much generosity is an example to the world.' But she was struck by the appearance of the arriving children: although their stomachs were empty they were swollen like balloons due to acute vitamin deficiency, brought on by worms.

A few days after the arrival of the landless untouchables, the peasants themselves came to seek refuge in Bhopal. The Kumar family, originally from a small village on the Indore road, had eight children. All of them had swollen stomachs, except Sunil, who at twelve was the eldest. Tales of this kind of famine were part of everyday life in India. Rice was invariably the protagonist: the rice they had planted, then lovingly pricked out; the rice they had caressed, palpated, sounded; the emerald green rice that had soon turned the colour of verdigris, then yellow for want of water; the rice that had drooped, shrivelled up, dried out and finally died. Nearly all the residents of the Kali Grounds were former peasants. Almost all of them had been through the same tragedy as these refugees who had sought asylum amongst them.

For the giant factory that stood a few yards away, this exodus was a bad omen. Warren Woomer's hopes were not to be fulfilled. Ten years earlier, Eduardo Muñoz had tried to make Carbide's directors appreciate a fundamental aspect of Indian existence: the vagaries of the monsoon. The people to whom the Argentinian spoke, however, had swept aside his warnings and responded with a figure. To a pesticide manufacturer, India meant half a billion customers! In the light of India's economic crisis at the beginning of 1983, that figure was now meaningless.

The failure of the publicity campaign Muñoz had launched compounded these unfavourable conditions. In vain Carbide flooded the countryside with posters depicting a Sikh holding a packet of Sevin and explaining to a peasant: *My role is to teach you how to make five rupees out of every rupee you spend on Sevin.* Farmers devoted most of their resources to buying seed and fertilizer. It was, moreover, more difficult than anticipated to induce peasants to change their traditional practices and adopt farming methods involving the intensive use of pesticides. Many farmers had come to realize that it was impossible to fight the onslaught of predatory insects in isolation. The insects migrated from treated areas to other untreated fields, before coming back to where they started as soon as the treatment that had driven them away had lost its efficacy. These frustrating comings and goings had contributed strongly to the decline in pesticide sales. In 1982, the Carbide representatives dispersed throughout the country had only been able to sell 2308 tons of their white powder. That was less than half the production capacity of the industrial gem designed by the ambitious young men of South Charleston. The forecasts for 1983 were even more pessimistic.

⁂

While stormclouds gathered over the future of the proud plant, something happened, one day, in a hut in Orya Bustee, that was to change Padmini's life completely. One morning, when she awoke on the *charpoy* she shared with her parents and brother, she found a bloodstain on her panties. She had started her first period. For a young girl in India this intimate progression is a momentous occasion. It means that she is ready for the one great event in her life: marriage. Custom may have it that a girl is married while she is still a child, but that is only a formality. The real union only takes place after puberty. Like all the other little girls of her age, even those from the humblest Adivasi families, Padmini had been prepared for the solemn day in which she would be the focus of everyone's attention. From her early childhood in Mudilapa and subsequently in Bhopal, she had learned everything that a good wife and mother of a family should know. As for her parents, they knew that they would be judged on the manner in which their daughter conducted herself in her husband's home, but her conduct would not be assessed exclusively on submission. Among the matriarchal Adivasis, women enjoy certain prerogatives usually reserved for men. One of them is that of finding a husband for their daughters. They are, however, spared the main task associated with this responsibility – that of gathering together an acceptable dowry – because it is the fiancé who brings his betrothed a dowry.

196

The daughter of an unskilled labourer, even one employed by Union Carbide, was not the most glittering match. Finding a husband would therefore take some time. But, as tradition required, that morning Padmini exchanged her child's skirt and blouse for her first sari. There was no celebration at the Nadars'. Her mother simply wrapped the panties in a sheet of newspaper. 'When we celebrate your marriage, we will go and take these to the Narmada,' she informed her daughter. 'We'll offer them to the sacred river in order that it may bless you and bring you fertility.'

❀

It is a well-known fact that love is blind. Especially when the object of one's passion is an industrial monster like a chemical plant. Warren Woomer had always refused to accept that the fate of Carbide's factory in Bhopal should be determined on profitability alone. No capitalist enterprise, however, could go on absorbing the loss of millions of dollars. The projections drawn up seven years earlier, predicting annual profits of seven to eight million dollars, were no longer remotely feasible. Could Woomer's replacement reverse the situation? The son of a former governor of the Reserve Bank of India, whose signature still appeared on two rupee notes, Jagannathan Mukund had many strings to his bow, but he was not a magician. As an undergraduate at Cambridge, he had been a brilliant chemistry student. He went on to complete his doctorate at MIT in Boston and was promptly snapped up by Carbide. He had spent two years in Texas, then two more running the petrochemical plant on the island of Trombay, and finally three years in Institute in West Virginia, mastering the complicated techniques involved in the production of MIC. Mukund was married to the daughter of the Deputy-Secretary of the

United Nations, herself a distinguished economist and university professor. He was the father of a little girl who was born with a deformity of the heart, whom American surgeons had saved by an operation which, at the time, they alone could perform. In theory, putting such an experienced man in charge of the factory was another gift from Carbide to its jewel in Bhopal. But it was in theory only. For the directors of the multinational's subsidiary were quick to subject the new Managing Director to a financial controller whose one and only purpose was to reduce the factory's losses, whatever the cost.

Cultured, refined and always supremely elegant in suits tailored in London, this 'superdirector', an aristocratic Bengali by the name of D. N. Chakravarty, was fifty-two years old. A great lover of poetry, high living, Scotch whisky and pretty women, he was certainly a distinguished chemist, but utterly unsuited to working in a plant that produced dangerous chemical substances. His entire career had been spent at the head of an industry where the worst that could possibly happen was that a conveyor belt broke down. The battery division he had run had in fact been a sinecure, reaping colossal profits without any risk whatsoever. The appointment of this intractable administrator would prove to be a fatal mistake.

# 29

## 'My beautiful factory was losing its soul'

The young engineer who had risked his life conveying the first barrels of MIC from Bombay to Bhopal could not get over it. 'When we were asked to show the new "superdirector" round the factory, it felt like taking a tourist round Disneyland,' Kamal Pareek recalled.

Chakravarty knew nothing at all about how a plant of that kind worked. He did not know what most of the components were for. He mixed them up: what he was calling a mixer was in fact a blender. In English the two words may mean essentially the same thing, but in Bhopal's technical jargon, they referred to different parts. 'We realized at once that this saviour they'd sprung on us was not party to the mystique of the chemical industry,' Pareek added. 'The only thing he was interested in was figures and accounts.'

This might still have turned out for the good, if only the new superdirector had been prepared to admit that a plant like that could not be run like a battery factory; if he could have graciously acknowledged that, in a company of that kind, decisions must come from all levels, each one affecting as it did the lives of thousands of people; if he had understood that seemingly favourable conditions could suddenly swing the other way, that the levels in the tanks were constantly rising and falling, that the combustion of the reactors varied by the moment; in short, that it was impossible to run that kind of plant simply by sending out memos from his directorial armchair. 'When you're in charge of a pesticide plant,' Pareek said, 'you have occasionally to come out of your office, put on overalls and join the workers on site, breathing in the smell of grass and boiled cabbage.'

Carbide's great achievement had been that of integrating a vast spectrum of different cultures and guaranteeing the humblest of its workers the right to speak. Unfortunately, neither Jagannathan Mukund, Woomer's successor, though steeped in considerable American experience, nor his superior from Calcutta, seemed inclined to engage in dialogue. Their understanding of human relations appeared to be based upon a concept of caste, not in the religious sense, but in a hierarchical sense. The introduction of such rifts was, little by little, to corrupt, divide and de-motivate. 'Once drastic cuts became the sole policy objective, and one man's say-so was the only authority, we knew the plant was inevitably going to hell,' Kamal Pareek maintained.

※

Once again it was Rahul who was the bearer of the news. In a matter of minutes it was all round the bustees.

'Carbide has just laid off three hundred coolies. And apparently that's only the beginning.'

'Haven't the unions done anything about it?' Ganga Ram asked sharply.

'They weren't given any choice,' explained Rahul.

'Does that mean they're going to shut down all the installations?' worried Sheela Nadar, who was afraid that her husband might be among the men laid off.

'Not necessarily,' Rahul tried to be reassuring. 'But it does seem the sale of plant medicines isn't going all that well any more.'

'It's not surprising,' observed Belram Mukkadam, 'the rains didn't come this year and people are leaving the country-side.'

Sunil, the eldest son of the Kumar family, they who had just left their rice fields obliterated by the drought, spoke then.

'Plant medicines are great when things are going well,' he declared. 'But when there's no water left to give the rice a drink, they're useless.'

Sunil was right. The gathering around the plank on wheels had increased in size. The news Rahul had brought pro-voked widespread consternation. After living so long in the shadow of the factory, after burning so many incense sticks to get jobs there, after being woken with a start by the howl of its sirens; after so many years of living together on this patch of land, how could they really believe that this temple of industry was crumbling?

'This year the rains are going to be very heavy,' Nilamber, whose predictions were always optimistic, reassured them. 'Then Carbide will take back those it kicked out today.'

Sheela Nadar gave the little man with the goatee beard a grateful smile. Everyone noticed that her daughter Padmini was wearing a cotton sari instead of her children's clothes.

'The trainees from the plant have stopped coming to the House of Hope,' Padmini added, alluding to the training centre Carbide had set up in part of the building occupied by Sister Felicity's handicapped children. 'The classrooms have been closed for several days. I don't think anyone's coming back because they've taken away all their equip-ment.'

Once again a shadow of desolation passed over all their

201

faces. Each one was silently contemplating the mighty structure looming on the horizon.

'I tell you they've only sacked our men so they can get even fatter off of them,' decreed Prema Bai, who had come from helping a new citizen of Orya Bustee into the world. 'Don't you worry: Carbide will always be there.'

※

The whole city adopted her opinion. Neither the death of one of its workers, nor the ensuing union unrest, nor the apocalyptic predictions of Rajkumar Keswani had been able to tarnish the factory's prestige in the Bhopalis' eyes. 'The Star' that Eduardo Muñoz and a group of impassioned engineers had constructed, belonged as much to the patrimony of the City of the Begums as did its mosques, palaces and gardens. It was the crowning glory of an industrial culture that was completely new to India. The residents of Bhopal might not always know what exactly the chimneys, tanks and pipework were for, but they enthusiastically joined in all the activities, sporting or cultural, organized by the plant. There were some indications, however, at the beginning of 1983, that the honeymoon period was drawing to a close. Under pressure from the top directors of Union Carbide, Chakravarty and Mukund devoted all their energies to making further cuts. 'In India, like anywhere else in the world, the only way to reduce expenditure is to reduce running costs,' Kamal Pareek said. 'In Bhopal, wages constituted the primary expense.' After the 300 coolies whose dismissal had upset the families in the bustees, it was the turn of over 200 skilled workers and technicians to be shown the door. Crucial though it was, in the methyl isocyanate production unit alone the manpower in each shift was cut by half. In the vitally important control room, only one man

202

# A miraculous pesticide to the rescue of the world's farmers

This sales' representative from the American multinational Union Carbide presents a box of Sevin to a poor farmer in Bengal, whose crops are regularly consumed by insects. 'Every rupee you spend on Sevin will make you five,' he assures him. This promise was to become Union Carbide's slogan to sell its product to India's four hundred million farmers.

It was in this research centre on the outskirts of New York City that, in 1957, two young entomologists and a chemist discovered Sevin, a miraculous, environmentally-friendly pesticide that was to save man's crops.

It was by using a gas called methyl isocyanate, that Union Carbide manufactured Sevin in its Institute plant in West Virginia, before producing it in Bhopal. Of all the gases the chemical industry has created, methyl isocyanate (MIC) is probably the most dangerous. This special steel tank at the Institute plant (behind Dominique Lapierre) could, in the event of an accident, release enough gas to kill the thousands of people living in the Kanawha Valley. It was from a very similar tank that, on the night of 2 December 1984, a toxic cloud emerged, killing between 16,000 and 30,000 people in Bhopal. As a result of that tragedy, safety systems have been tightened considerably.

1

2

3

4

Since the dawn of time, men have been waging a relentless war against the small creatures that attack their crops.

1. The asparagus beetle
2. The Colorado potato beetle
3. The black bean aphid
4. The colza flea beetle
5. The beetroot weevil
6. The chestnut beetle

# America's gift to India: a pesticide

Bill Sneath (fourth from the left), then Chairman of Union Carbide, the world's third largest chemical giant, lands in the heartland of India on 19 January 1976 with a group of engineers to inaugurate the first work units of the plant that aimed to free Indian farmers from the scourge of devouring insects.

The Chairman's hand is on a box of Sevin. This contains a hundred one-hundred gram packages of insecticide mixed with sand or gypsum powder, the product was to be spread by hand or sprayed by airplane over the large estates in the Punjab.

# plant built by its best engineers

A marriage of West and East; the then Chairman of Union Carbide, Bill Sneath, with his wife, between the Moghul-style colonnades of a terrace overlooking the Bhopal Upper Lake.

An Indian engineer submits samples of the product that is to protect Indian crops to Union Carbide's then Chairman, Bill Sneath.

With its sublime mosques, magnificent palaces and superb gardens, the city of Bhopal (population six hundred thousand) was, for centuries, a highpoint of India's culture. Its many marvels earned it the nickname 'India's Baghdad'.

Until India's independence in 1947, princes and princesses reigned over the Muslim kingdom of Bhopal. The last nawab, Hamidullah Khan, used to parade amongst his subjects on the back of an elephant.

Four generations of enlightened begums made the state of Bhopal one of the most
modern in Asia. Completely concealed behind her *burkah*, the begum Shah Jahan (front
row, centre) poses outside her palace with representatives of British imperial power. She
had sent her son to fight in the trenches of the Great War and gave all her gold to the
Allies' cause.

The begum Sikander returns to her carriage after inaugurating the railway
station that would put an end to Central India's isolation. She had financed
the railway with her own funds. She also instituted compulsory public
education.

# They had all dreamed of lighting

Union Carbide entrusted the building of its Bhopal plant to the elite of its engineers. For each one of them, the venture was an extraordinary crusade in which all kinds of cultures, religions and rituals mingled under the aegis of chemistry.

Eduardo Muñoz. This Argentinean agronomic engineer had sold Sevin all over South America. With its 400 million peasants, India represented a huge potential market for him.

Warren Woomer. A specialist in high-risk chemical installations, this American engineer imposed very strict safety measures during the two years he headed the Bhopal plant.

John Luke Couvaras. Given the different cultural backgrounds of his skilled and unskilled workers, he regarded the construction of the Bhopal plant as a remarkable achievement.

Kamal Pareek. This Indian engineer tried to instill a safety culture into all the plant's staff. When he discovered that economic cuts were jeopardizing the safety of the plant, he handed in his resignation.

Mohan Lal Varma. This was the workman Union Carbide accused of causing the catastrophe by deliberately introducing water into tank 610. His alleged sabotage was never proved.

Shekil Qureshi. For this Indian technician, there was no more beautiful uniform than Carbide's boiler suit. He risked his life trying to prevent the disaster and was the last to leave the plant.

This is the 'beautiful plant' from which, on the night of 2 December 1984, forty-two tons of methyl isocyanate escaped. The wind was blowing in the direction of the neighbouring slums. The toxic cloud caused between 16,000 and 30,000 deaths. Months before, for reasons of economy, all the safety systems had been deactivated.

Under the effects of the heat caused by the chemical reaction, tank E610 burst out of its concrete sarcophagus, but it did not shatter. Today it lies there amongst the weeds of the abandoned plant.

Amongst the debris from the factory, Javier Moro found the temperature gauge for tank E610. The instrument was not working on the night of the tragedy. No one realized that a gas explosion was going to occur.

# A wreath of hovels under the very walls of the murderous factory

Drawn by the jobs Union Carbide provided, thousands of poverty-stricken people crammed themselves into a wreath of hovels surrounding the factory. Every morning, children scavenged the railway tracks for the pitiful pieces of treasure fallen from the trains, and sold them on to rag-pickers. Women begged a little hot water from the engine drivers.

The former leprosy sufferer Ganga Ram and his crippled wife Dalima were one of the most popular couples in Orya Bustee. Ganga Ram gave the slum its first television set. This brought about a real revolution in the residents' lives, who discovered that another world existed beyond the walls of their shacks.

Colonel Khanuja, the Sikh commander of the Bhopal Engineer Corps unit, was one of the heroes of the fatal night. With nothing to protect him but fireman's goggles and a wet handkerchief over his face, he sprung at the head of a column of trucks to rescue four hundred workers of a cardboard factory and their families who had been surprised by the gas in their sleep.

# These were the actors and victims in the worst industrial disaster in history

Sajda Bano. Two years before the catastrophe, her husband was the first victim of the plant's gas. On 2 December 1984, Sajda was traveling back to Bhopal by train with her two children. The deadly vapours that had invaded the station immediately killed her eldest son.

Reverend Timothy Wankhede. Vicar of the Church of the Holy Redeemer, he had spent his Sunday visiting the sick in the city's hospitals. Thinking an atomic bomb had fallen on Bhopal, Timothy took his wife and his son to the altar of his church, opened the book of the Gospel according to St Matthew and recited: 'Ye know not what hour the Lord doth come.' A moment later, he and his family were miraculously saved.

Rajkumar Keswani. A journalist, Keswani predicted the catastrophe in four articles published two years before it happened. But this far-sighted Cassandra was a voice crying in the wilderness. No one took his predictions seriously.

Deputy Station Master V. K. Sherma. He saved several hundred lives when he ordered the engine driver of a train full of pilgrims who had come to the city to celebrate Ishtema, to leave the Bhopal railway station. Believed dead from gas inhalation, V. K. Sherma was taken to the morgue, only to be rescued at the last minute from the funeral pyre. Today, he still suffers from the physical and psychological consequences of the tragedy.

No one will ever know precisely how many people perished in the course of that tragic night and the following days. Dr Satpathy, pathologist, had the photos of a large number of the victims put on display. Four hundred of them were never claimed: whole families had been wiped out and many of the dead had not had a fixed address.

Dr Deepak Gandhe was on duty at Hamidia Hospital on the night of the catastrophe. He was the one who first attended to the victims before being swamped by the tide of dying people. He remained at his post for three days and three nights.

# 6,000 and 30,000 thousand lives

By the early hours of the morning of 3 December, Hamidia Hospital had become one gigantic charnel house. Muslims were buried in communal graves and truckloads of wood had to be brought from all over the region to burn the Hindus. Most of the victims had died of cardiac and respiratory arrest.

Respiratory problems, persistent coughs, vision impairments, early-age cataracts, anorexia, recurrent fevers, weakness, depression, cancers, tuberculosis – the survivors are still suffering from the after effects of the tragedy.

For Union Carbide's victims, calling for the death of Warren Anderson, chairman of the company at the time of the catastrophe, is meagre consolation. But the vengeful graffiti on the walls of Bhopal are a reminder that a court has yet to try and punish those responsible for the tragedy.

While the dead were laid out in their hundreds under the walls of Hamidia Hospital (above), survivors wait for treatment in the tents of an improvised hospital. Their eyes burnt by the gases, many of them would never recover their sight (below). Hundreds of doctors from all over India tried everything to save these wretched people. Because Union Carbide had never revealed the composition of the toxic gas, no antidote was readily available.

# The massacre of the innocents

This father carried his child in his arms to Dr Gandhe, crying out to him: 'Save him!' 'Alas he's dead,' responded the doctor. The poor man refused to believe it. He ran off, leaving the child in the arms of the man from whom he expected the impossible.

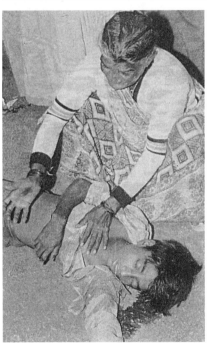

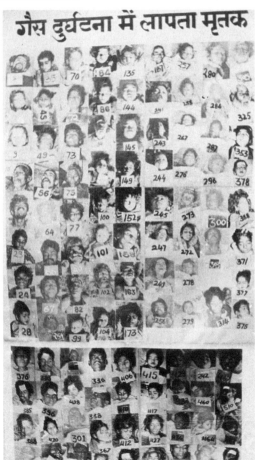

गैस दुर्घटना में लापता मृतक

A mother with her son who had been struck down and a page from a tragic album: two images of a city martyred by the madness of an industrial giant. In order to save just a few hundred rupees, all the factory's safety systems had been deactivated.

was left to oversee the some seventy dials, counters and gauges, which relayed, among other things, the temperature and pressure of the three tanks containing the MIC. Maintenance crews underwent the same cuts. Out of a total of nearly 1000 employees, the plant soon had no more than 642 left. What was more, 150 workers found themselves withdrawn from their regular workstations to make up a 'pool' of manpower that could be moved here and there as the need arose. The result was a drop in the standard of work as many specialists found themselves assigned to tasks for which they had not been trained. The replacement of retiring skilled personnel with unskilled workers made further savings possible, at the risk of having key positions filled by inexperienced people. The latter often spoke only Hindi, while the instruction manuals were written in English, which only added to the confusion.

Kamal Pareek would never forget 'the painful meetings during which section heads were obliged to present their plans for cuts.' The most senior engineers were reluctant to suggest solutions that would compromise the safety of their installations. But the pressures were too great, especially when they came from the Danbury head office in the United States. That was how the decision was reached that certain parts, which were supposed to be changed every six months, would be replaced only once a year. And to replace any damaged stainless steel pipes with ordinary steel piping. Numerous cuts were made along those lines. Chakravarty, the man primarily responsible for this flurry of cutbacks, seemed to know only one metal, the tin-plate used in batteries. He behaved as if he knew nothing about corrosion and the wear and tear on equipment subject to extreme temperatures. 'In a matter of weeks, I saw everything I'd learned on the banks of the Kanawha River, go out the window,' Pareek concluded. 'My beautiful plant was losing its soul.'

Pareek had put his faith in the new values preached by the multinational. But in reality that magnificent edifice, he was suddenly discovering, was founded on one religion alone: the religion of profit. The blue and white hexagon was not a symbol of progress; it was just a commercial logo.

⁂

No ceremony was held to mark the departure of D. N. Chakravarty in June 1983. He left Bhopal, satisfied that he had been able in part to stem the factory's haemorrhaging of finances.

Jagannathan Mukund was left in charge, but with a mission to continue the policy of cutbacks initiated by the envoy from Calcutta. He rarely left the air-conditioned ivory tower of his office and took six months to reply to the three inspectors from South Charleston who had reported multiple breaches of the safety regulations. He assured them that 'all defects would be duly corrected', but the time schedule he proposed for restoring standards left them mystified. Some of the faulty valves in the phosgene and MIC units would not be able to be replaced for several months. As for the automatic fire detection system in the carbon monoxide production unit, it could not be installed for a year at the earliest. These grave infractions of the sacrosanct safety principles were to soon produce a new cry of alarm from the journalist Rajkumar Keswani. The factory was continuing to go downhill. The maintenance men had no replacement valves, clamps, flanges, rivets, bolts or even nuts. They were reduced to replacing defective gauges with substandard instruments. Small leaks from the circuits were not stopped until they were really dangerous. Many of the maintenance procedures were gradually phased out. Quality control checks on the substances produced became less and less frequent, as did the checks on the most sensitive equipment.

'Produce when required' . . . Soon the factory only went into operation when the sales team needed supplies of Sevin. This was precisely the method that Eduardo Muñoz had tried, ten years earlier, to convince the engineers in South Charleston to adopt, to avoid stocking the enormous quantities of MIC required for continuous production. Now that the plant was operating at a reduced pace, Mukund stopped MIC production in order gradually to empty the tanks. Soon they held only about forty tons. It was a trivial quantity by Institute 2 American standards but enough, if there were an accident, to fulfil Rajkumar Keswani's apocalyptic predictions.

In the autumn of 1983, Mukund made a decision which was to have far-reaching consequences. Ignoring his predecessor's warning, he ordered the shutting down of the principal safety systems. In his view, because the factory was no longer active, these systems were no longer needed. No accident could occur in an installation that was not operating. His reasoning failed to take into account the forty tons of methyl isocyanate sitting in the tanks. Interrupting the refrigeration of these tanks might possibly save a few hundred rupees worth of electricity a day, and possibly the same amount in freon gas. But it violated a fundamental rule laid down by Carbide's chemists, which stipulated that methyl isocyanate must in all circumstances be kept at a temperature close to 0° celsius. In Bhopal, the temperature never drops below 15° or 20° celsius, even in winter. Furthermore, in order to save a few hundred pounds of coal, the flame that burned day and night at the top of the flare was extinguished. In the event of an accident this flame would burn off at altitude any toxic gases that spilled into the atmosphere. Other pieces of essential equipment were subsequently deactivated, in particular the enormous scrubber cylinder, which was supposed to decontaminate any gas leaks in a bath of caustic soda.

⁕

There were many engineers who were unable to bear the degradation of the high tech temple they had watched being built. Half of them had left the factory by the end of 1983. On 13 December, it was time for the one who had been there longest to go. For the man who had so often risked his life escorting trucks full of MIC from Bombay to Bhopal, it was a departure that was both heart-rending and liberating.

Before leaving his beautiful factory, Kamal Pareek wanted to show his comrades that in case of danger the safety systems so imprudently shut down could be started up again. Like a sailor climbing to the top of his ship's main mast to light the signal lamp, he scaled the ladder to the immense flare and relit the flame. Then he headed for the three tanks containing the methyl isocyanate and unbolted the valves that supplied the freon to the coils that kept them refrigerated. He waited for the needle of the temperature gauge to drop back down to 0° celsius. Turning then to K. D. Ballal, the duty engineer for the unit that night, he gave a military salute and announced:

'Temperature is at zero celsius, Sir! Goodbye and good luck! Now let me run to my farewell party!'

# 30

## *The fiancés of the Orya Bustee*

'D on't cry, my friend. I'll take you to the Bihari's place. He already has a drove of about a hundred. He might like to take on one more.'

After Belram Mukkadam, Satish Lal, a thin, bent, good-natured little man with protruding muscles, was one of the longest-standing occupants of the slum. He lived in the hut opposite Padmini's family. He had left his village in Orissa to find work in the city, in order to pay back the debts he had incurred for his father's cremation. A childhood friend, who had come back to the village for the festival of Durga, had enticed him to Bhopal where he was a porter at the main station. 'Come with me,' he had said, 'I'll get you a coolie badge and you'll buy yourself a uniform. You'll make fifteen to twenty rupees a day.' So Satish Lal had worked at Bhopal station for thirty years. His seniority gave him a certain prestige in the porters' union, which was led by a man from the state of Bihar who was known simply as the Bihari. Now Satish Lal hoped that his standing as a trade unionist would enable him to help his neighbour, Ratna Nadar, Padmini's father, who had been laid off by Carbide, along with 300 other unskilled workers.

'You never actually see the Bihari,' explained Satish Lal. 'No one even knows where he dosses down. He's a gang leader. He couldn't give a damn whether it's you or Indira Gandhi carrying the luggage along the platforms, just so long as every evening you pay him his whack, in other words a share of your tips. An employee of his takes care of that. He's the only one who can get you the badge authorizing you to work as a coolie. But don't think he's any easier to approach than his boss. You have to be introduced to him by someone he trusts. Someone who'll tell him who you are, where you come from, what caste, what line of descendants, what clan you belong to. And it's in your interests to greet him with your sweetest *namaste* and throw in plenty of *sardarjis*[1], as many as ever you like. And invoke upon his person the blessing of Jagannath and all the deities.'

'Shouldn't I also give him something?' asked the former Carbide worker anxiously. He had always had to grease the *tharagars*' palms to get himself taken on.

'You're right my friend! You'd better have fifty rupees ready, some *pan* and a good dozen packets of *bidis*. Once he's accepted you, the police will look into whether you've been in any trouble in the past. There again you'd better have some backsheesh to hand.' Ratna Nadar's eyes widened as the sum he would have to lay out grew. – 'And then there's the stationmaster's PA. He passes on the green light from the police to his boss, who is the guy who gives you your badge. Your badge is a talisman. When your bones ache too much for you to carry bags and suitcases, you can pass it on to your son. But be careful, if you refuse to take a minister's or some other big shot's baggage because they never tip, the stationmaster can take it away from you.'

In the time he had been working at the station, Satish Lal

---

[1] Sardar: chief; *sardarji*: term of respect expressed in the suffix ji.

had done it all. He even claimed to have carried on his head the enormous trunks of Hamdullah Khan, the last Nawab. They were very heavy; the locks on them were solid silver.

With his *lunghi* bulging at the waist with rupees for the various intermediaries, Nadar set out for the station in the company of his neighbour. Before entering the small office next to the cloakroom occupied by the coolies' union, the two men stopped before an altar, which harboured, beside a tulsi tree, an orange statue of the god Ganesh. Ratna Nadar rang the small bell in front of the divinity to ask for his protection and placed a banana and a few jasmine petals in the offering bowl.

Ganesh fulfilled Padmini's father's wishes. A few days later, his neighbour burst into his hut.

'You did it, my friend!' he announced triumphantly. 'You're Bhopal station's one hundredth and first coolie. Go quickly and buy yourself a red tunic and turban. And a supply of titbits and sweets. The stationmaster's waiting to give you your badge.'

❋

It was a ritual. At each full moon, the elders of the Kali Grounds would take their places on sisal mats laid end to end, men on one side, women on the other, to discuss the affairs of their community. The men would exchange *pan* and *bidis*, the women sweets. One of the purposes of these meetings was to review the young people of the neighbourhood who had reached marriageable age. Their names were listed and a debate at once ensued. Soon certain boys' and certain girls' names would find themselves linked together. Comment on the merits and disadvantages of these hypothetical marriages would redouble. So seriously did the inhabitants of the bustees take their family lineage, that the process was sometimes carried over to the next meeting. One day Prema Bai spoke up.

'We have to find a good husband for Padmini,' she said emphatically.

'Prema Bai's right,' said the lovely Dalima.

There followed some discussion. Several boys were mentioned, among them Dilip, Dalima's adopted son. For that reason Dalima followed the conversation with rapt attention. As usual, Belram Mukkadam tried to calm things down.

'There's no rush,' he declared. 'As I understand it, Padmini Nadar is still too young.'

'You've been misinformed, brother,' the girl's mother immediately replied, 'she's reached marriageable age. And we want to find the best possible husband for her.'

'You couldn't find a better husband for your daughter than my son, Dilip,' Dalima said proudly. 'He's an exceptional boy and I want a wife for him who is no less so.'

The real meaning of this statement was lost to no one. Its purpose was less to extol the boy's virtues than it was to make certain that Sheela's expectations regarding dowry were realistic.

'My daughter is just as exceptional as your son,' advanced Sheela. 'And if your son is such a treasure, you will of course have anticipated giving him a generous dowry.'

'I had anticipated doing my duty,' responded Dalima, anxious to avoid confrontation at this stage in negotiations.

The discussion went on within the framework of a very precise ritual, which neither of the two parties could breach. It would take two more assemblies under the full moon and a lot of debate to reach agreement over the union of Dilip and Padmini. The transaction could then proceed to the *manguni*, the official request for the girl in marriage. Out of respect for tradition, the boy's parents invited several of the neighbourhood's elders to represent them in this traditional formality. But, as always in India, no ceremony could take place without first consulting a *jyotiji*, an astrologer, who was to examine the stars to see whether the proposed couple were compatible and determine the most propitious date for the *manguni*. In the neighbouring Chola Bustee lived an old man with a white beard, called Joga who, for forty years, had been a fortune-teller on the streets of the old city of Bhopal. His was not always an easy task, especially when, as with Dalima and the Nadars, the parents of the prospective marriage partners did not know the exact date on which their children had been born. Old Joga confined himself to suggesting that the marriage request should take place during a month under the benign influence of the planet Venus, and on a day of the week which was not Friday, Saturday or Sunday, the three inauspicious days of the Indian lunar-solar calendar.

⁂

A procession as elaborate as that of a royal ceremony came to a halt outside the Nadars' hut. In the recollection of Orya Bustee, there had never before been such a *manguni*. Ganga Ram had arranged for a kid goat to be cooked, and the elders accompanying him arrived with their arms full of victuals, sweets, and bottles of beer and country liquor.

It was a real *barakanna*[2], a celebration such as the occupants of the bustees had never previously known. Ganga Ram, who had conquered leprosy, put his crippled wife back on her feet and given the community a television to enable them to dream along with India's epic tales every evening, had also shown himself to be the most generous of stepfathers. On behalf of her daughter, Padmini's mother

---

[2] Literally: 'great banquet'.

211

accepted the *pindhuni*, the silk outfit decorated with gold thread that he brought as an official and tangible expression of the promise of matrimony. The engaged couple did not take part in this ceremony. All the preparations for their marriage went on without them. As is customary they would not meet each other until the wedding night, when, as a symbol of their marriage, Dilip would lift the veil from his fiancée's face to place red *sindur* powder on the parting of her hair. However, Dilip and Padmini had clearly known each other for a long time.

With the banquet over, the time came to move on to the most serious issue, the dowry. It was to old Prema Bai that Padmini's mother had entrusted the role of negotiating this important ritual. With the help of some of the other women, she had drawn up a list of the items and gifts Dilip's family would be expected to give his future wife. The list included two cotton saris, two blouses, a shawl and various household utensils. It also included jewels: some imitation, others real, in this instance two rings, a nose stud and a *matthika*[3]. As for gifts for the bride's family, they were to include two *dhotis* for her father, two vests and two *punjabis*, the long tunic buttoned from the neck to the knees. Her mother was to receive two silk saris and a pair of sandals encrusted with small ornamental stones. They were poor people's requirements, certainly, but they were worth some 3000 rupees, a fabulous sum even for the proprietor of a small painting firm.

Belram Mukkadam, Iqbal and Rahul, who represented Dilip's family, had listened to the croaky voice of the elderly midwife without flinching. Marriage negotiations being traditionally long-winded affairs, custom had it that the groom's family clan consulted together before giving its consent.

---

[3] Ornament worn on the forehead.

Dalima was so keen for her son to marry Padmini, however, that the three envoys wagged their heads at the same time, indicating that they accepted all the girl's family's conditions.

It was then that old Joga, the white-bearded astrologer, who had silently witnessed all these exchanges, threw caution abruptly aside.

'Before you conclude your haggling, I would appreciate it if you would agree the remuneration for my services,' he declared vehemently.

'We thought of two *dhotis* for you and a sari for your wife,' replied Mukkadam.

'Two *dhotis* and a sari!' exclaimed the *jyotiji*, beside himself. 'You've got to be joking!'

From the recesses of the huts, the entire alleyway followed this unexpected turn-up with avid interest.

'If you're not satisfied, we'll find another *jyotiji*,' threatened Rahul.

The astrologer burst out laughing.

'I'm the one who drew up the horoscopes! No one else will agree to choose the marriage date instead of me!'

This reply was greeted with much chortling from the onlookers. Some of the women heckled: 'He's a real son of a bitch, that *jyotiji*!' sneered one of them. 'More than that, he's devious,' replied another. Suddenly, Dalima's voice erupted like thunder. Her beautiful green eyes were bloodshot. She was fuming.

'You piece of shit!' she shouted. 'If you wreck my boy's marriage, I'll skin you alive!'

The astrologer made as if to get up and go. Iqbal held him by the arm.

'Stay,' he begged.

'Only if you pay me a hundred rupees deposit straight away.'

The participants looked at each other helplessly. All of a

sudden, however, there was the stocky figure of Ganga Ram. He was holding a bundle of notes between the stumps of his right hand.

'There you are,' he said dryly, dropping the notes into the little man's lap. 'Now tell us on what day we should celebrate our children's marriage.'

The astrologer went through the motions of thinking. He had already done his calculations. He had eliminated all the days when the sun entered the ninth and twelfth signs of the zodiac, and chosen one when the sun was favourable for the groom while the planet Jupiter was most beneficent for the bride.

'The second of December, between ten o'clock and midnight, will be the most propitious time for your children's union,' he announced.

# 31

## *The end of a young Indian's dream*

T he document was stamped BUSINESS CONFIDEN-
TIAL and dated 11 September 1984. Addressed to the
person in charge of Union Carbide's engineering and safety
department in South Charleston, it was signed J. M. Poulson,
the name of the engineer who, two years previously, had
headed the safety audit of the Bhopal factory. This time
Poulson and the five members of his team had just finished
inspecting the storage conditions of several hundred tons of
methyl isocyanate in the Institute plant, right in the heart of
the Kanawha Valley, home to more than 250,000 Americans.

The document revealed that this American installation was
suffering from a number of defects and malfunctions: vibration likely to rupture sensitive piping; potentially dangerous
leakage from various pumps and other apparatus; corrosion
of electric cable sheathing; poor positioning of several automatic fire extinguishers in sectors of prime importance; faults
in the filling systems to the MIC tanks etc. In short, deficiencies that proved that safety at the flagship factory left a
lot to be desired. The document also claimed that the actual
health of personnel working in Institute was at risk. Poulson
and his team had in fact discovered that workers in the MIC

unit were often subjected to chloroform vapours, especially during maintenance operations. There was no monitoring system to measure the duration of their exposure, despite the fact that chloroform was a highly carcinogenic substance. The report stipulated that an interval of fifteen minutes would constitute dangerous overexposure. All the same, the investigators considered these risks relatively minor by comparison with the danger 'of an uncontrollable exothermic reaction in one of the MIC tanks and of the response to this situation not being rapid or effective enough to prevent a catastrophe'. The document gave a detailed list of the circumstances which made such a tragedy possible. The fact that the tanks were used for prolonged storage was conducive to internal contamination that was likely to pass unnoticed until precisely such a sudden and devastating chemical reaction occurred. The investigators had actually found that the tanks' refrigeration system introduced minuscule impurities, which could become the catalysts for such a reaction. They had discovered that these impurities could also come from the flare meant to burn off the toxic gases at a height of 120 feet. In short, the most modern plant, the safest in the whole of the United States' chemical industry, appeared to be at the mercy of a few drops of water or metal filings. 'The potential hazard leads the team to conclude that a real potential for a serious incident exists,' ended the document. In his accompanying letter, Poulson gave the names of sixteen Carbide executives who should receive copies of his report. Strangely, this list made no mention of the man to whom it was a matter of primary concern. Jagannathan Mukund, Managing Director of the Bhopal plant, with three tanks permanently holding forty tons of MIC, would remain ignorant of the concerns expressed by the American engineers and, in particular, of their recommendations to counteract a possible accident.

The plant on the Kali Grounds was a little like his baby. It was he who had set down the plans for the first formulation unit. It was he who had bought the splendid palace from the Nawab's brother, to turn it into an agronomical research centre. Together with Eduardo Muñoz and several other fanatical pioneers, Ranjit Dutta had laid the foundations of the 'beautiful plant' right in the heart of the City of the Begums. As far as this engineer with the physique of a rugby player was concerned, his time spent in Bhopal had been a magical period in a richly successful career. After leaving India in 1976 to take over the running of Carbide's American operations, Dutta had repeatedly returned to the site of his first love. Every year he went there on holiday with his family, to boat on the waters of the Upper Lake, listen to poets sing on evenings when there was a *mushaira* in Spices Square, and dream beside the illuminated outline of the vessel whose funnels he had designed.[1]

Now, at the age of fifty-four, he was Vice-President of the agricultural products division at the company's head office. And that summer of 1984, at the time when the team of investigators led by Poulson were compiling their report, the Indian engineer had just come back from a pilgrimage to Bhopal. This time, however, the man who loved the city so much came back sad and disappointed. 'I didn't like what I saw during that visit,' he recounted sorrowfully. 'I saw the approaches to the factory overrun with rubbish and weeds. I saw unoccupied workers chatting for hours over cups of tea. I saw mountains of files strewn about the management's

---

[1] Although originally from the region that was to become Bangladesh and despite having spent part of his life in the United States, Europe and the largest cities of India, it was to Bhopal that Ranjit Dutta returned to retire.

offices. I saw pieces of dismantled equipment lying about the place. I saw disorientated, unmotivated people. Even if the factory had temporarily stopped production, everyone should have been at their workstations getting on with maintenance work . . . It's strange but I sensed an atmosphere of neglect.'

As soon as he got back to Danbury, Dutta tried to relay this impression to his superiors but, oddly, none of them wanted to listen. 'They probably thought I was harbouring some sort of grievance against the local management,' he said, 'or that I wanted to take over the running of the factory again. But I only wanted to warn them that strange things were going on in Bhopal, and that people there were not doing their jobs as they should.'

It would not be long before Dutta had an explanation for this indifference. If no one at the top of Union Carbide seemed interested in the 'neglect' to which the factory had fallen prey, it was for a good reason: in Danbury the Bhopal plant had already been written off. Dutta would have formal confirmation of the fact at the conference, which, every year, assembled the heads of the company's agricultural divisions in the Connecticut countryside. At this meeting, specially adapted marketing strategies for products made by Carbide throughout the world – sales prices, methods of beating the competition and acquiring new clients – were fixed. There the fate of the Bhopal factory had already been discussed. As early as 1979, the economic viability of the plant had been subject to extensive debate. One of the various options considered was simply stopping its construction. Because of the stage the building work had reached, this idea had been abandoned. Five years later, the situation had further deteriorated. The plant was now losing millions of dollars. The sales prospects for Sevin in 1984 did not exceed 1000 tons, half the amount for the preceding year and only a fifth of its total production capacity. It was a disaster. Approval was given for a liquidation programme. In fact, the multinational was counting on getting rid of its costly Indian factory by moving its installations to other Third World countries. Brazil, for example, could accommodate the phosgene, carbon monoxide and methyl isocyanate units. As for the Sevin formulation and packaging works, Indonesia seemed the ideal place for them to be relocated.

Carbide's Vice-President for Asia sent a top secret message to Bhopal. He wanted to know the financial and practical feasibility of dismantling and moving the plant, 'taking into account the moderate price of Indian labour'.

The task of gathering the necessary information was entrusted to the Hindu engineer Umesh Nanda. Nine years earlier, a brief advertisement in the *Times of India* had enabled this son of a modest industrialist in the Punjab to fulfil the dream of all young Indian scientists of his generation: that of joining a renowned multinational. Now, he was charged with shattering his own dream. Dismantling and dispatching the Sevin production unit by ship should not pose any problem, he responded in a telex to his superiors on 10 November. The same would not appear to be true of the MIC unit, however, because of extensive corrosion damage. Nanda warned that the reassembly of this unit could only be undertaken after repair work involving considerable expense. The Indian's telex provided the answers to Carbide's queries. It also confirmed what had been Rajkumar Keswani's worst fears. The 'beautiful plant' had been abandoned.

After a two-year absence, Rajkumar Keswani had just arrived back in Bhopal. He was not yet aware that Carbide had decided to write the factory off and was preparing to transfer it to another Third World country. Ever more disquieting intelligence from his informants inside the plant induced him to sound a fourth alarm, entitled BHOPAL ON THE BRINK OF DISASTER. This time he really believed that his article would rouse public opinion and convince the authorities. *Jansatta*, the regional daily that ran his pieces, was not a local journal but one of India's biggest newspapers, belonging to the prestigious *Indian Express* group. Once more, however, Keswani was to be a voice crying in the wilderness. His latest apocalyptic predictions provoked not the slightest interest in the public, any more than they incited the municipal authorities to take any safety measures. The journalist sought an explanation for this latest failure. Wasn't I convincing enough? he asked himself. Do we live in a society where people mistrust those interested in the public good? Or do they just think I'm putting pressure on Carbide to fill my own pockets?

The wheel of destiny was turning. In a few weeks' time, Keswani's round face would appear on all the world's television screens and he would become the youngest reporter ever to receive the Press Award of India, the highest possible distinction accorded to a journalist in India.

# 32

## *The vengeance of the people of the Kali Grounds*

Not for the world would she have missed her meeting with the ordinary people of India. Every morning before leaving to perform her onerous duties as Prime Minister of the world's most populous democracy, Indira Gandhi received those who came to seek a *darshan*, a face-to-face communication, with the woman who embodied supreme authority. The encounter took place in the rose and bougainvillaea-laden garden of her residence on Safdarjang Road, in New Delhi. For the 67-year-old patrician who these seventeen years had ruled over a fifth of all humanity, such morning gatherings were an opportunity to immerse herself in the multifaceted reality of her country. Draped in a sari, she would move from group to group, speaking for instance to peasants from the extreme south; then to a delegation of railway workers from Bengal; then to a group of young schoolgirls with long plaits; and thereafter to a squad of barefoot sweepers who had come from their distant province of Bihar. To each one the mother of the nation had a few words to say: she read the petitions presented to her, responded with a promise and posed graciously for souvenir photographs. Thus, as in the days of the Mogul emperors, remotest parts of India had, for a moment's interlude, daily access to the seat of power.

That morning of Wednesday 31 October 1984 promised to be a splendidly clear, bright autumn day. A soft breeze rustled the leaves of the neem trees in the vast garden where the lucky few waited to receive their morning *darshan*. They were joined by a British television crew who had come to interview her. On the previous evening, Indira had returned from an exhausting electoral tour of Orissa, the native state of most of the refugees in Orya Bustee. In the presence of the thousands of followers who had come to hear her, she had concluded her speech with surprising words, 'I don't have the ambition to live a long life, but I am proud to live it serving the nation. If I were to die today, each drop of my blood would make India stronger.'

At eight minutes past nine, she walked down the three steps from her residence into the garden. She was wearing an orange sari, one of the three colours of the national flag. On passing the two sentries on either side of the path, she pressed her hands together at the level of her heart in a cordial *namaste*. The two men had the traditional Sikh beards and turbans. One of them, 40-year-old Beant Singh, was well known to her: for ten years he had formed part of her closest bodyguard. The other, 21-year-old Satwant Singh, had had only four months service. A few weeks previously, Ashwini Kumar, former Director General of the Border Security Force of India, had come to see Indira Gandhi to express his concern. 'Madam, do not keep Sikhs in your security service.' He had reminded her that Sikh extremists had sworn to pay her back for the army's bombardment and bloody seizure of their most sacred sanctuary, the Golden Temple of Amritsar. On the preceding 6 June, the attack had killed 650 Sikhs. Indira Gandhi had smiled and reassured her visitor. Indicating the figure of Beant Singh in the garden, she had replied: 'While I'm fortunate enough to have Sikhs like him about me, I have nothing to fear.' Sceptical, the former Director General had insisted. She had brought their meeting to a close with some irritation: 'How can we claim to be secular if we go communal?'

On that 31 October she had scarcely finished greeting the two guards when the elder pulled out his P38 and fired three bullets point blank into her chest. His young accomplice promptly emptied the magazine of his Sten gun into her body. At least seven shots punctured her abdomen, ten her chest, several her heart. The mother of India did not even have time to cry out. She died on the spot.

❊

Just as the assassination of Mahatma Gandhi thirty-six years previously had done, the news plunged the nation into a state of painful stupor. By the middle of the afternoon, every city in India had become a ghost town. In Bhopal, a twelve-day period of mourning was decreed. All ceremonies, cele-brations and festivities were cancelled, while cinemas, schools, offices and businesses closed their doors. Flags were flown at halfmast. Newspapers brought out special editions, in which they invited readers to express their despair. 'India has been orphaned,' proclaimed one of them. Another wrote: 'In a country as diversified as ours, only Indira could guar-antee our unity.'

'We shall no longer hear the irresistible music of her elo-quence . . .' In Bhopal, people recalled with sadness her recent visit for the inauguration of the Arts and Culture building. 'The realization of this project will make Bhopal the cultural capital of the country,' she had announced to applause and cheers of *Indira Ki Jai!* The city's companies, businesses and organizations filled the newspapers with notices expressing their grief and offering their condolences. One of these messages was signed Union Carbide, whose entire staff, so it maintained, wept for the death of India's Prime Minister.

That afternoon, the shattered voice of the governor of Madhya Pradesh resounded over the airwaves of *All India Radio*. 'The light that guided us has gone out,' he declared. 'Let us pray God to grant us the strength to remain united in this time of crisis.' A little later the local people gathered round the transistor belonging to Jai Prakash's bicycle repairman. Arjun Singh, Chief Minister of Madhya Pradesh, who had made them property owners by granting them their *patta*, was also expressing his sorrow. 'She was the hope of millions of poor people in this country,' he declared. 'Whether they were Adivasis, *harijans*, inhabitants of the bustees or rickshaw pullers, she always had time for them and a solution to offer to their problems. [. . .] May her sacrifice inspire us to continue to go forward . . .'

It was only on the next day, however, when the funeral was held in New Delhi, that the residents of Bhopal along with all the people of India really became conscious of the tragedy that had befallen their country. For the first time in history, television was going to broadcast the event as far afield as the most remote of the continent's countless villages, to wherever a television set belonging to some *zamindar*[1], privileged person, organization or club, could show the images relayed live from India's capital. All at once an entire nation was to be joined together by media communion. At daybreak, at the behest of Ganga Ram, owner of the only television set in the area, the Kali Grounds' huts were empty of all occupants. Belram Mukkadam and Iqbal

---

[1] Large landowner.

had stacked several of the tea house tables on top of one another and covered them with a large white sheet, a symbol of purity and mourning, and then decorated it with garlands of marigolds and jasmine flowers. Then they had positioned the set high enough for everyone to see the screen.

Since the early hours of the morning, the crowd had been gathering in silence outside the tea house: men on one side, women and children on the other. They watched silently as representatives of the country's different religions succeeded one another, reciting prayers and appealing for forgiveness and tolerance.

Suddenly, a murmur rose from the assembly. Wide-eyed, the residents of the Kali Grounds began to witness the transportation to the funeral pyre of the woman who, only the previous day, had ruled their country. The litter, covered with a bed of rose petals, jasmine flowers and garlands of marigolds, filled the screen. Indira Gandhi's face, with the veil of her red cotton sari set like a halo around it, emerged from an ocean of flowers. With her eyes closed and her features relaxed, she radiated an unusual serenity. The screen showed hundreds of thousands of Indians massed along the funeral route, leading to the sacred banks of the Yamuna River where the cremation would take place. The cameras lingered on tearful faces, on people clinging to street lamps and branches of trees, or perched on rooftops. Like waters coming together again in the wake of a ship, the crowd rushed in behind the funeral carriage -- ministers, coolies, office workers, businessmen: Hindus, Muslims and even Sikhs in their turbans – representatives of all the castes, religions, races and colours of India, all united in shared grief. For three hours this endless river swelled with fresh waves of humanity. When, finally, it reached the place where a pyre had been built on a brick platform, to the Kali Grounds' residents it looked as if a groundswell suddenly surged through the hundreds of thousands of people gathered there. To Padmini, all those people looked like millions of ants in a nest. To old Prema Bai, who remembered seeing photographs of Mahatma Gandhi's funeral, it was the finest tribute to any servant of India since the death of the nation's liberator. Amongst the crowd of television viewers, a woman with short hair said her rosary: Sister Felicity was intent upon sharing the sorrow of her brothers and sisters in the bustees.

As soon as the funeral carriage stopped, a squad of soldiers lifted the mortal remains of Indira Gandhi and carried them to the pyre. Television viewers saw a man dressed in white, with the legendary white cap of the Congress Party and a white shawl lined in red over his shoulders. They all recognized Rajiv Gandhi, Indira's elder son, her heir and the man the country had chosen to succeed her. Traditionally, it was his responsibility to carry out the last rites. The cameras showed him spreading a mixture of ghee, coconut milk, camphor essence and ritual powders over his mother's corpse. While the television set flooded the tea house esplanade with *Vedic* mantras recited by a group of priests in saffron robes, Rajiv took hold of the cup containing the sacrificial fire. Five times India's new leader circled the pyre, from left to right, the direction in which the Earth revolves round the Sun. The crowd saw his son, Rahul, appear next to him, together with his wife Sonia and their daughter Priyanka. Although women usually did not take part in cremations, Rajiv's wife and daughter helped to place firewood round the body. A camera focused next on the flaming cup, which Rajiv raised for a moment above the surrounding heads before plunging it into the pyre. When the first flames began to lick at the blocks of sandalwood, a voice intoned the same prayer that Belram Mukkadam had recited on the death of his father:

226

*Lead me from the unreal to the real,*
*From darkness to light*
*From death to immortality . . .*

At once a mighty howl issued from the crowd assembled around the Kali Grounds' tea house.

The cry uttered over 600 miles away acted like a detonator. Suddenly, the voice of Rahul drowned out the sound of the television. 'We should avenge Indira!' he yelled. His usually smiling mouth was twisted with fury. 'Rahul is right, Indira should be avenged!' numerous other voices took up the cry. 'This city's full of Sikhs. Let's go and burn down their houses!' shouted someone. At this cry, everyone got up, ready to rush to Hamidia Road and the area round Bhopal's main *gurdwara*[2]. Climbing onto the platform, Ganga Ram addressed the multitude.

'No need to go to Hamidia. It would be enough . . .'

He had no time to finish his sentence. Ratna Nadar had jumped on the platform.

'Friends, Nilamber has just been found hanging from a beam of his hut. On his *charpoy*, there is a picture of Indira and a garland of flowers.'

Nilamber, the sorcerer whom everybody loved because he only ever predicted good fortune! The news of his suicide bewildered all those present. Death was a familiar event here but this time it was different. Nilamber had been overcome by grief. It was Belram Mukkadam's turn to mount the stage.

'Ganga's right,' he cried. 'It isn't worth going all the way to Hamidia Road to set fire to the Sikh houses, it would be enough to set fire to the usurer's, the man who sucks us dry. Everyone to Pulpul Singh's house!'

By setting fire to Pulpul Singh's house they would be

---

[2] Sikh temple.

making a Sikh pay for the horrible murder perpetrated by two of his brothers in religion, but they would also be avenging all the crimes committed by the money lender, who had, at one time or another, humiliated each and every one of them. His safe already contained several property deeds mortgaged against pitiful loans. Pulpul Singh was the ideal scapegoat. By setting fire to his house, obliging him to flee, perhaps even killing him, they would be avenging Indira, avenging Nilamber, avenging all the injustices of life.

At the first cry for vengeance a woman has slipped away from the crowd. Her God was a God of love and forgiveness. For her the word 'vengeance' did not exist. Her duty was to prevent her brothers' and sisters' anger ending in tragedy. Spotting the dark silhouette hurrying away, Padmini joined her. Preempting her question, Sister Felicity took the young Indian girl by the arm and swept her along with her.

'Come with me quickly to Pulpul Singh's. We must warn him so he has time to get away.'

The Sikh usurer's residence was a solid two-storey house built at the entrance to Chola. A veranda protected by railings ran along the entire façade of the ground floor. It was from behind the shelter of these bars that Pulpul Singh conducted his trade, enthroned like a Buddha on a velvet stool in between his safe and a portrait of Guru Nanak, founder of his religion. Behind him hung a chromo of the Golden Temple of Amritsar, the destruction of which had provoked Indira Gandhi's assassination.

The usurer was surprised by the arrival of the two women. Neither the nun nor young Padmini belonged to his usual clientele.

'What wind of good fortune blows you this way?' he asked.

'Get out of here! For the love of God, leave immediately with your family!' the nun begged him. 'They want to take vengeance on you for Indira Gandhi's assassination.'

She had scarcely finished speaking when the frontrunners of the crowd arrived. They were armed with iron bars, pickaxes, bricks, bolts and even Molotov cocktails. 'For the first time I saw a sentiment on their faces that I had thought not to find in the poor,' Sister Felicity would recall. 'I saw hatred. The women were amongst the most overwrought. I recognized some whose children I'd nursed, even though their contorted features made them almost unrecognizable. The residents of the Kali Grounds had lost all reason. I realized then what might happen one day if the poor from here were to march on the rich quarters of New Bhopal.'

Terrorized, Pulpul Singh and his family fled out of the back of their house, but not before wasting precious time trying to push their safe to the back of the veranda and hide it with a cloth. In the meantime, the rioters had thrown their first bottle of flaming petrol. It hit the ground just behind Sister Felicity and Padmini who had remained outside. The explosion was so powerful that they were thrown towards each other. Dense smoke enveloped them. When the cloud cleared, they found themselves in the middle of the rampaging crowd. The shoemaker Iqbal had brought a crowbar, to force open the railings. Suddenly someone shouted: 'Get them! They've escaped out the back!' A group took off in pursuit of the fugitives. Their Ambassador had failed to start, so they were trying to get away on foot. Restricted by their saris, the women had difficulty running. Soon the family was caught and brought roughly back to the house. In his flight Pulpul Singh had lost his turban.

'We're going to kill you,' declared Ganga Ram, caressing the man's throat with the point of a dagger. 'You're scum. All Sikhs are scum. They killed our Indira. You're going to pay for that.'

With his shoulder, he shoved the usurer up against the bars on the terrace.

'And you can open up your shit hole of a house at once, otherwise we'll set fire to it and you.'

Terrified, the Sikh took a key from his waist and unlocked the padlock to the grille. Cowering together, Sister Felicity and Padmini observed the scene. The nun recalled something an old man from Orya Bustee had explained to her one day: 'You keep your head down, you wear yourself to a frazzle, you put up with everything, you bottle up your bitterness against the factory that's poisoning your well, the usurer who's bleeding you dry, the speculators who are pushing up the price of rice, the neighbours' kids who stop you sleeping by spewing up their lungs all night, the political parties that suck up to you and do damn all, the bosses that refuse you work, the astrologer who asks you for a hundred rupees to tell you whether your daughter can get married. You put up with the mud, the shit, the stench, the heat, the mosquitoes, the rats and the hunger. And then one day, bang! You find some pretext and the opportunity's given you to shout, destroy, hit back. It's stronger than you are: you go for it!' Sister Felicity had often marvelled that in such conditions, there were not more frequent and more murderous outbreaks of violence. How many times in the alleyways had she seen potentially bloody confrontations suddenly defused into streams of verbal insults and invective, as if everyone wanted to avoid the worst.

A series of explosions shook the Sikh's house. Immediately afterwards the veranda went up in flames. There were shouts of: 'Death to Pulpul Singh!' And others of: 'We're avenging you, Indira!' Salar appeared brandishing a knife. 'Prepare to die!' he shouted, and advanced towards the terror-crazed Sikh. Another second and Salar would have lunged at the usurer. But the moment he raised his arm, someone intervened.

'Put down your knife, brother,' ordered Sister Felicity, seizing the young man firmly by the wrist.

230

Stunned, Salar's friends did not dare interfere. Ganga Ram stepped forward, accompanied by his wife Dalima. Although she walked unsteadily, she had managed to catch up with the crowd. She had just seen the nun position herself between Salar and the usurer.

'Killing that bastard wouldn't do any good!' she cried, turning on the rioters. 'I've a better idea!' With these words she took from the waist of her sari a small pair of scissors. 'Let's cut this Sikh's beard off! That's a far worse form of vengeance than death!'

Ganga flashed his wife a smile of admiration.

'Dalima's right, let's cut the shit's beard off and throw it on the flames of his house.'

Salar, the tailor Bassi and Iqbal grabbed hold of the usurer and pinned him to the trunk of a palm tree. Dalima handed the scissors to Belram Mukkadam. After all, it was only right that the manager of the tea house should have the honour of humiliating the man who had exploited him for so many years. Resigned to his fate, the usurer did not protest. The process took a while. Everyone held their breath. The scene was both pathetic and sublime. When there was not a trace of hair left on the usurer's cheeks, neck or skull, a joyful ovation went up into a sky obscured by the smoke from his flaming house. Then Mukkadam's deep voice was heard to say:

'Indira, rest in peace! The poor of the Kali Grounds have avenged you.'

❊

The vengeance wrought by the occupants of the slums on the Sikh usurer was only the tiniest spark in a terrible explosion that erupted throughout India against the followers of Guru Nanak. The flames of Indira Gandhi's funeral pyre had scarcely gone out before violence, murder and pillage was unleashed in the capital and the country's principal cities. Everywhere Sikhs were brutally attacked; their houses, schools and temples were set on fire. Soon fire brigades, hospitals and emergency services were overwhelmed by the flare up of violence, that reminded many people of the worst horrors of Partition in 1947. Despite a rigorous curfew and the intervention of the army, more than three thousand Sikhs were immolated on the altar of vengeance.

On the morning of 2 November this violence hit the City of the Begums in a particularly horrible fashion. Colonel Gurcharan Singh Khanuja, the Sikh officer in command of the electrical and mechanical engineer corps stationed in Bhopal, came out of his barracks accompanied by an escort to go to the railway station. Several members of his family – his two brothers, his brother-in-law and several nephews – were returning from a pilgrimage to the Golden Temple of Amritsar. When Khanuja opened the door to the compartment reserved for his family, he found nothing but charred corpses. Assassins had stopped the train as it was leaving Amritsar, slit the throats of all the Sikh passengers and set fire to their corpses.

※

Five days later, a special train decorated with flags and garlands of flowers pulled in at the same platform, no 1, in Bhopal station. It was bringing the population one of the thirty-two urns containing the ashes of the dead Prime Minister making their way round the country. A guard of honour made up of soldiers in full uniform, their rifle butts inverted, and a brass band playing the funeral march, waited to take the precious relic to an altar erected in the middle of the parade ground where the city's poetry evenings were usually held.

The entire town gathered along the route. Belram Mukkadam, Ganga Ram, Dalima and Dilip, Padmini and her parents, Salar, all the occupants of the Kali Grounds, including Prema Bai and Rahul on his plank on wheels, were there with the rest of the population to pay their respects to the woman who had proclaimed that the eradication of poverty should be India's first priority. For two days thousands of Bhopalis of all castes, religions and origins came to throw flowers at the foot of the altar decorated with the flags of the country and of Madhya Pradesh. Banners identified these endless processions. There were groups from the Congress party, associations, clubs, representatives of all conceivable organizations and creeds.

After its sojourn in Bhopal, followed by a pilgrimage through various other cities of Madhya Pradesh, the urn was taken back to New Delhi, from where a military aircraft escorted by two Mig 23s transported it, with all the other urns, over the highest peaks of the Himalayas. On board the airplane was Rajiv Gandhi. He emptied all the ashes into a basket, which he covered with a red satin veil. When the plane was flying over the eternal snows of the river Ganges' birthplace, India's new leader cast the basket into the crystal clear air. Indira Gandhi's ashes were scattered, to return to the high valleys of Kashmir, the land of the gods and the cradle of her family.

# 33

## *Festivities that set hearts ablaze*

November, the month for festivities. While Union Carbide abandoned its Indian industrial masterpiece to its sad fate, the unconcerned City of the Begums gave itself up to all the joy and celebration of the world's most festive calendar. Nowhere did this taste for rejoicing manifest itself with as much intensity as in the Kali Grounds' bustees. There, festivals wrested the poor from the harsh realities of their lot. A more effective vehicle for religion than any catechism, they set hearts and senses ablaze with the charm of their songs and the rituals of their long and sumptuous ceremonies.

The Hindus opened the festivities with a frenzied four-day festival in honour of Durga, the goddess conqueror of the buffalo demon that rampaged through the world a hundred thousand years ago. The entire city was filled with splendid *pandals*, temporary altars, to hold the statues of the goddess, all dressed up and magnificently bejewelled. Two such altars brightened up the otherwise gloomy environs of the Chola and Jai Prakash bustees. For four days, people processed past them, regardless of any distinctions of faith. The men wore woollen *sherwanis* over their trousers; the women silk *kurtas* and dangling earrings that made them look like royalty.

At twilight on the fourth day, the statues of the goddess were hoisted onto a luggage cart that Ratna Nadar had borrowed from the station. His wife and Dalima had draped it with a piece of shimmering cloth and decorated it with flowers. Ganga Ram's musicians were there again to accompany the procession. At the same time, in other parts of Bhopal, other similar corteges were setting out. They made for the shores of the Upper Lake in the heart of the city, where the statues crowned with their gilded diadems were immersed in the sacred waters, bearing with them all the joys and afflictions of the Bhopalis.

A little while later, it was the Muslims' turn to celebrate the anniversary of the birth of the Prophet Mohammed. The Kali Grounds' families painted their homes, outside and in, with whitewash tinged with green, the colour of Islam. Chains of multi-coloured bunting were strung across the alleyways. Prostrate in the direction of the mystical and distant Kaaba, Salar, Bassi, and Iqbal spent a night of devotion, squeezed with hundreds of other faithful into the two small mosques built beside the railway line in Chola and Jai Prakash. Next day a human tide, vibrant with faith and reciting *suras* at the tops of their voices, poured through the neighbourhood's alleyways. '*Allah Akbar!* God is great!' recited the multitude from beneath banners representing the domes of the sacred mosques of Jerusalem, Medina and Mecca, symbols that imbued the place with faith, piety and fantasy.

The Muslims had barely finished commemorating the birth of Mohammed before a myriad luminous snakes streaked across the sky above the Kali Grounds. Celebrated during one of the longest nights of the year, Diwali, the Hindu festival of lights, marks the official arrival of winter and symbolizes the triumph over darkness. The illuminations were to celebrate one of the most beautiful episodes in the *Ramayana*, the return of the goddess Sita to the arms of her divine husband Rama after her abduction by the demon Ravana. That night in their huts, Hindu families played cards like mad, for the festival also commemorated the famous dice game, in which the god Shiva won back the fortune he had lost to Parvati, his unfaithful wife. To achieve this victory, Shiva appealed to his divine colleague Vishnu, who very opportunely assumed the form of a pair of dice. Diwali was thus a homage to luck. The residents gambled with ten, five or one rupee notes, or even with small coins. The poorest would gamble a banana, a handful of puffed rice or some sweets. Every alleyway had its big gambler. Often it was a woman. The most compulsive was Sheela Nadar. Padmini would look on bewildered, as her mother shamelessly fleeced old Prema Bai.

'It's a good omen, my girl!' Sheela would explain after every winning hand. 'The god of luck is with us. You can rest assured that your marriage will be as beautiful an occasion as Diwali.'

In exactly one week's time, on Sunday 2 December, the happy conjunction of Jupiter and the Sun would transform Padmini into a princess out of *A Thousand and One Nights*. On that day, Jagannath, the glorious avatar of Vishnu worshipped by the Adivasis from Orissa, would bless her marriage to Dilip.

The ritual for an Adivasi marriage is as strict as any uniting high caste Hindus. Nine days before the ceremony, Padmini and Dilip had to submit themselves to all kinds of ablutions in the homes of neighbourhood families, before a meal and

236

the presentation of gifts to equip their household. Four days later, the married women took charge of the young couple for a purification ceremony, in which they were rubbed down with castor oil and other ointments smelling strongly of saffron and musk. Once this oiling had been completed, they proceeded to the couple's toilet and the interminable trying on of the wedding outfits made by Bassi. The cost of these outfits had been subject to keen negotiation. For a humble coolie working at Bhopal station, marrying his daughter off meant substantial sacrifices.

Three days before the wedding, Ratna Nadar and several of his neighbours built the *mandap*, the platform on which the union would be celebrated. This was a dais about ten yards wide, raised about twenty inches from the ground and made out of mud coated with a smooth, dry mixture of cow dung and clay. Branches from two of India's seven sacred trees covered the sides of the platform and, in the middle, on an altar decorated with flowers, stood the image of Jagannath. Strings of light bulbs provided the finishing touch to the decorations. On the evening of the ceremony, they would be lit by a generator hired for the occasion. Belram Mukkadam had chosen a prime position. Padmini and Dilip would be married on the esplanade outside the tea house, where all the community's great events took place, looking out at the tanks and pipework of the plant which, to all those poor people, represented the hope of a better life.

# 34

## A Sunday unlike any other

The dawn prayer. Every morning Bhopal awoke to the call of the muezzins from high up in the minarets. That Sunday 2 December 1984, however, was no ordinary day. In a few hours' time the City of the Begums was due to celebrate Ishtema, the great prayer gathering which, once a year, brought thousands of pilgrims from all over the country as well as Pakistan and Afghanistan, to the heartland of India. Ratna Nadar had been obliged to temporarily abandon preparations for his daughter's wedding and go with the other station porters to meet the special trains overflowing with the faithful. There would never be more people in Bhopal than on that Sunday. The excitement had already come to a head in the Taj ul-Masajid, the great mosque where teams of electricians were installing the floodlighting which would illuminate the splendid building for a week. Volunteers were unrolling hundreds of prayer mats and connecting up loudspeakers which, for three days and three nights, would sound out prayers in celebration of the greatness of Allah.

All round the city's mosques and outside the hotels on Hamidia Road, the bus station and the railway station, hundreds of street vendors were taking up their positions. Ishtema was a lucrative time for any business in Bhopal. Jolly and rubicund, his lip accentuated by a thin moustache and his forehead decorated with Vishnu's trident, Shyam Babu, a 45-year-old Hindu, was the proprietor of the city's largest restaurant. Muslim, Hindu or secular, the many festivals in the Indian calendar made his fortune. Situated in the old part of the city, his establishment, the Agarwal Poori Bhandar, could serve up to 800 settings a day and never closed. 'Our meals are the best and the cheapest in town,' he assured people. And it was true that for ten rupees, the equivalent of less than thirty pence, one could eat one's fill of vegetables, chicken or fish curry and samosas. But Shyam Babu was not just a businessman; he was also a kind man. The lepers and beggars who hauled themselves up the steps of the great mosque, and the penniless pilgrims who camped out in the ruins of the palace of Shah Jahan Begum, knew that they would always find a bowl of rice and vegetables if they went to him.

Shyam had started that Sunday as he began every day, with a morning prayer in the small temple to Lakshmi, goddess of wealth. He had taken her baskets of fruit and flowers, for he was going to have particular need of her support that day. For him, the eve of any festival was always difficult. The massive arrival of visitors meant that many police reinforcements had to be brought in. The municipal government counted on Shyam to feed these men. It had become a tradition. The restaurateur had ordered up an extra 650 pounds of potatoes, the same quantity of flour, and doubled the stocks of fuel to supply his fifteen ovens. 'Don't you worry, I could feed the whole city,' he informed the police chief who had come to make sure that his men would be adequately nourished.

Not far from Shyam Babu's restaurant, a noticeboard drew attention to another, rather quaint business, which had sprung into action for this Sunday unlike any other. For three generations the Bhopal Tent and Glass Store had been hiring out equipment and accessories for the city's weddings and public celebrations. The grandson of its founder, Mahmoud Parvez, a Muslim who looked like a mullah with his little goatee beard and his embroidered skullcap, ran his business by telephone from a work-table set up in the street. The warehouse behind him was a veritable Ali Baba's cave whose secrets he alone knew. In it were piles of plates, crates of glassware, drawers full of cutlery, candlesticks of all sizes, old gramophones, antique generator sets, elephant bells, flintlock guns and arquebuses. Parvez's pride and joy was a gleaming Italian percolator. 'I'm the only one in town, in the whole of Madhya Pradesh even, who can serve espresso coffee!' he boasted. What had earned him most renown, however, was his impressive collection of carpets and *shamianas*, the multi-coloured tents used for public and private ceremonies. He had them to suit all tastes and all purses. Some of them could hold up to 2000 guests. Others, by virtue of their age and the refinement of their patterns, were real museum pieces. Parvez only hired them out on very special occasions and then only to friends or people of prominence. That Sunday, his staff were laying on the most wonderful *shamiana* for the wedding of the daughter of the controller-in-chief of the Bhopal railway, Ashwini Diwedi, whose brother Sharda was Managing Director of the city's power station, two people of standing whom Mahmoud was eager to please. The remaining rugs and *shamianas* were for the day's numerous other weddings and the Ishtema on the following day, not forgetting the *mushaira*, the poetry recital arranged for ten o'clock that night in Spices Square. For this event, Parvez would also be providing small cushions so that the poets could relax between recitations, accessories all the more necessary because a number of these men of letters were members of the celebrated Lazy Poets' Circle.

Mahmoud Parvez rubbed his hands as he watched his storehouse empty. That Sunday was going to be an auspicious one for the Bhopal Tent and Glass Store.

❄

The feverish preparations had spread as far as the Kali Grounds' two main artisans. Mohammed Iqbal had been working since dawn to finish shoes made out of Agra leather and sandals encrusted with precious stones ordered by several guests attending the day's weddings. With the help of his young apprentice, Sunil Kumar, the son of poor peasants newly arrived in the bustee, he cut, trimmed and sewed away, surrounded by the suffocating smell of glue and varnish that filled the hut where his wife and three children were still sleeping. Across the way, in hut number 240, his friend Ahmed Bassi had also been up since dawn, finishing embroidering the saris and veils ordered by the wealthy families of Arera Colony for their daughters' weddings. Bassi had such fine silk fabrics brought from Benares that his shop attracted Bhopal's smart set, despite its location in the poor quarter. Five times a day, he thanked God for all the benefits He had bestowed upon him. His order book was overflowing. In two weeks' time, it would be Eid, the most important festival in the Muslim year. The treadle of his sewing machine would not stop, as he made *kurtas* out of satin and *sherwanis* in Lucknow brocade.

❄

At the other end of town, in a church with a slate-covered steeple in the Jehangirabad district, on that same 2 December, the small minority of Christians in Bhopal has assembled to celebrate one of the major events in the liturgical calendar. The first Sunday in Advent was the beginning of a time of prayer and recollection leading up to the anniversary of the birth of Jesus, who came to redeem humanity's sins. There was a Christmas crib with life-size figures to commemorate the birth of the Messiah on the straw of a stable in Bethlehem. A noisy and colourful congregation, with women in superb saris, the embroidered ends covering their heads, and sumptuously dressed men and children, filled the nave, cooled by a battery of fans. Majestic in his immaculate alb and red silk vestments, Eugene de Souza, the Roman Catholic Archbishop, originally from Goa, read the first psalm with fervour: 'Awake thy glory, O Lord, and deliver us, for our transgressions have led us into imminent danger.'

That morning one pew remained unoccupied. Sister Felicity had called the prelate to ask him to excuse her and to request that his vicar, Brother Lulu, come to Ashanitekan, the House of Hope, to say Mass for the handicapped children in the small chapel there. To the right of the altar was a large picture of Jesus, under which were inscribed the simple words: 'I am with you always.'

A dozen children were kneeling on jute sacks sewn end to end. Among them was Nadia, the little girl with cerebral palsy, whom the nun had put in her own bedroom in order to better care for her. For much of the time, especially at night, her illness plunged her into a coma-like state, almost as if she were dead. The previous night, however, Nadia had suddenly woken up, screaming. 'People with this kind of illness have a very special sensitivity,' Sister Felicity explained. 'Nadia never woke up in the night unless something unusual was going to happen, like a storm, or the beginning of the monsoon. But the weather was so beautiful in Bhopal that second day of December that I couldn't understand why all at once she started to yell.'

The nun was to find her answer in the Gospel reading for the day: The sun shall be darkened, and the moon shall not give her light, and the stars shall fall from heaven, and the powers of the heavens shall be shaken . . .

✳

In the northern part of the immense city, near the central railway station, the Anglican incumbent of the small white church of the Holy Redeemer was also meditating with his flock upon the sombre predictions of the Holy Scriptures. Short, stocky with a round, smiling face, the 31-year old vicar Timothy Wankhede had come originally from Maharastra. Together with his wife and 10-month-old baby, Anuradh, meaning Joy, he lived in a modest red-brick vicarage next to the church. Like Archbishop de Souza, he poured endless energy into keeping the flame of Christian faith alight in a city inhabited by an overwhelming majority of Hindus and Muslims. Timothy had become a Christian one day while listening to the radio. He was twenty years old when an announcement in Marathi, his mother tongue, suddenly came over the airwaves. 'He who chooses and believes in Jesus Christ will be saved and all his kinfolk with him' was the message. 'I was overwhelmed,' Timothy recalled. 'I rushed to the only public telephone in the village and called the radio station, wanting to know more about Jesus Christ.' After being baptized, on 'the most wonderful day' of his life, he had travelled about India for three years, preaching the Gospel. Then he had spent four years at a theological college studying for the ordination that would open the doors of the parish of Bhopal to him.

The Reverend Wankhede's ministry was not confined to leading worship. That first Sunday in Advent he was preparing to take his parishioners to visit the city's various hospitals. 'It's our duty to comfort our suffering brethren,' he directed them, 'and tell them that Jesus' hands can heal, if only we believe in him.' In his shoulder bag he carried editions of the Bible in a dozen different languages. For that Advent Sunday he had chosen to read a verse from St Paul to the sick, which in a few hours time would prove to be tragically relevant. 'O God, forgive your children who were missed by those who had lured them with the promise of wealth.'

⁂

The two men were practitioners of a medical speciality of which crime writers are particularly fond. Sixty-two-year-old Professor Heeresh Chandra and his young assistant, Ashu Satpathy, performed autopsies on the corpses which sundry incidents throughout the year – accidents, crimes or suicides – dispatched to the examination tables of the Department of Pathology at the Gandhi Medical College. In a city with 600,000 inhabitants, there were plenty of violent deaths, even on Sundays and holidays. In the absence of a suitably refrigerated morgue, the two pathologists had to be constantly available to perform autopsies as soon as the corpses came in.

With his dignified air and imposing white moustache, Professor Chandra looked like a maharajah from the Rajput era. Even more than his exploits of cutting corpses, it was his hobbies that had earned him notoriety amongst the Bhopalis: he collected dogs and vintage cars. He owned three yellow Labradors and a 1930s National, known throughout the city for the way its exhaust backfired. That 2 December, the eccentric professor was getting ready to take his venerable vehicle and his Labradors out for a drive, as he did every Sunday, to Delawari nature reserve, a favourite resort of the Bhopalis.

His young colleague, Ashu Satpathy, who was always impeccably dressed in a tie, spent his leisure time indulging his passion for roses. Not having a large enough garden at his Idgah Hills cottage, he had transformed the corridors and terraces of the Department of Pathology into a rose garden. Dozens of containers and pots of flowers stood alongside the rows of jars containing the livers, kidneys, hearts, spleens and brains that enabled him to extract information from the bodies brought in by the police. Satpathy devoted any free time he had watering, pruning and feeding his dwarf bushes and climbing roses. The same fingers that immersed themselves in human entrails, carried out delicate grafts to produce new varieties, the secrets of which he alone knew. He had given them such lyrical names as 'Black Diamond', 'Moschata rose', 'Chinensis', 'Odorata', and 'Golden Chrysler'. In two days' time the doctor was going to exhibit all these wonders in the greenhouses at the monumental flower show which, for one week, would turn Bhopal into India's rose capital.

Alas the events of that Sunday were to thwart the two doctors' plans. Towards midday a telephone call from police headquarters informed them that two bodies, those of a man and a woman, were on their way to the morgue. It was a matter of urgency to establish the cause of death.

Before starting work, the two pathologists enlisted the help of the accomplice who was party to all their dissections. With his beige cap eternally crammed down over his long hair, the 28-year-old photographer Subhash Godane looked more like an artist than an accessory to a post mortem examination. He dreamed of making his mark on the world of fashion and advertising photography and had assembled

245

an impressive portfolio of women's portraits which he was preparing to show at the Delhi biennial exhibition. In the meantime his Pentax K-1000 supported his wife and three children by photographing corpses riddled with stab wounds, decapitated children and women who had been slashed to pieces. Godane was absolutely convinced that his camera had registered every conceivable horror humanity could inflict. He was wrong.

The autopsies on the two bodies took three hours. The absence of any traces of violence on the couple, who were both in their forties, suggested a double suicide by poisoning. Analysis of the internal organs confirmed Doctors Chandra and Satpathy's hypothesis. In the victims' stomachs they found copious quantities of a whitish powder that had caused extensive damage to the digestive and respiratory organs. The two practitioners were unable to determine the precise nature of the substance, but they were probably dealing with a strong pesticide in the DDT family. The police from Bhopal's criminal investigation department, who went to the village where the bodies had been found, had discovered that the victims were peasants whom the latest drought had reduced to ruin. Unable to pay back the loans they had taken out to buy seed, fertilizer and insecticides for their next crop, they had decided to end it all. Such cases were by no means unusual in India, nor was the method used. That Sunday 2 December, Carbide's beautiful factory had started to sow its seeds of death. In the two peasants' hut, the police found an empty packet of Sevin.

❄

A Sunday of prayer and mourning but a Sunday of folly too. Round a circle of dust in an old hangar attached to the

Lakshmi Talkies, the city's oldest and largest cinema, clustered three hundred over-excited punters. The building shook with all the shouting and heckling and the din from the loudspeakers. Men in shirts and *lunghis*, their fingers clutching bundles of rupees, pushed their way through the onlookers to pick up the bets. In the front row of the arena, an old man with a light complexion, dressed in a *kurta* of an elegance out of keeping with the general scruffiness, was silently massaging the claws of a cock. Omar Pasha, the godfather of the bustees, never talked before a fight.

Pressed round him like a bodyguard were his friends from the Kali Grounds led by Belram Mukkadam, Ganga Ram and Rahul. They had all bet on Yagu, Omar Pasha's champion, the creature with the murderous spurs he was holding on his belly and who, if he won that afternoon, would open the way to the championships in Ahmedabad in January, then Bangalore in March and finally New Delhi in April. Omar Pasha gently massaged the bird's thighs, joints and claws, while the creature relaxed, clucking with pleasure. Then, with the help of a file, he sharpened its spurs and beak into deadly daggers.

The sound of a gong announced the beginning of the fight. The godfather stood up and carefully placed Yagu in front of his opponent. The two cocks immediately hurled themselves at one another with a fury that roused the fever of their audience. Beaks and spurs spun in the light like steel-tipped arrows. The blood spurting in all directions did nothing to diminish the fury of the two combatants. The crowd yelled their names, clapped and stamped their feet. When one of the birds rolled over in the dust, the audience was delirious. Omar Pasha followed the ferocious set-to with the detachment of a Buddha. Yagu bled, staggered and fell but each time he got up to strike again. With a final blow of his spurs he managed to put out one of the eyes of his adversary, who collapsed, mortally wounded. Another sound of the gong signalled the end of the fight. The godfather stood up and retrieved his bloody but victorious cockerel. Parading the creature above his head like a trophy, he greeted the crowd.

# 35

## A night blessed by the stars

A Sunday of frivolity and freedom from care. Usually closed on Sundays, the stores in the Chowk Bazaar, scattered round the minarets and golden-spired cupolas of the Jami Masjid, were experiencing a record trade. That 2 December was, above all else, a day for marriages blessed by the stars. Elegant ladies from the smart neigbourhoods came rushing in to make last minute purchases. Necklaces, earrings, bracelets, all kinds of jewellery that were a speciality of Bhopal, were snatched up. Perfumers were robbed of their scent bottles of sandalwood, attar of roses and patchouli. Vendors were plundered of their silks, ribbons and sandals. It was as if the end of the world were at hand.

A splendid institution inherited from the British, the Arera Club on the other side of town was doing the sort of business it did on festival Sundays. Its members thronged round an abundantly laden buffet table, onto the tennis courts, into the Olympic-sized swimming pool and the reading rooms, and round the edges of the immaculately manicured lawns.

Executives from Carbide and Bhopal's other companies were entitled to membership of this club, nestling in an oasis of mauve, blood-red, orange and white bougainvillaeas, palm trees, frangipani and neem trees. With its gala evenings, balls, tennis and bridge tournaments, and games of bingo, the Arera Club had at one time given the South Charleston expatriates and their young Indian colleagues a glimpse of the life led by its British members in the great days of the Empire. Recently things had changed somewhat. On that Sunday 2 December 1984, there were no longer any American Carbiders standing round the pots of chicken curry and other Indian delicacies on the buffet. There were hardly even any Indian engineers left because the factory had been deserted by so many of its local senior staff. One of its few remaining representatives, Managing Director Jagannathan Mukund, had come to lunch with his youngest son who was in Bhopal for his university vacation. That evening, Mukund and his wife planned to take him to several marriage celebrations. And next day, they were going to show him some of the picturesque sites surrounding the City of the Begums. The plant had ceased operations so there was no reason its captain could not be gone for a day or two.

Not far from the Mukunds' table, a heated rubber of bridge was going on. One of the players was a young doctor in white trousers and a sports shirt. Both a swimming and a bridge champion, the athletic, 32-year-old doctor L. D. Loya had been recruited in March by classified advertisement to take over the running of the dispensary Carbide had built on the actual factory site. For the son of a Rajasthani corn chandler who had struggled hard to get his degree in toxicology, landing a job for an international company making chemical products was an achievement. In eight months, Loya had not had to deal with a single serious medical emergency, which was just as well because the management had not provided him with any detailed information about the composition of the principal and most

250

dangerous gas produced by the plant, and even less about how to treat the effects of it in case of accident.

❉

He was the man who probably would have the most onerous responsibilities on that remarkable Sunday. Sharda Prasad Diwedi was the Managing Director of Bhopal's power station. That evening, his turbines would have to supply enough current to light up the many feasts and wedding celebrations. The grandest was to take place in the Railway Colony. It was to mark the nuptials of Rinou, youngest daughter of the chief controller of the Bhopal railway.

The Railway Colony was typical of the neighbourhoods built by the British to house railway employees close to the stations in which they worked. A small town within a town, not too unlike the villages of Sussex or Surrey, with its lawns, its cottages, its cricket pitch, tea room, bank and a church with a Victorian gothic bell tower. With also one of those institutions, which seem to crop up whenever two Englishmen get together: a club. On that particular Sunday, the colonial-style railway employees' club accommodated the parents and at least two hundred guests of the groom's family. Later that evening, over a thousand people were due to squeeze themselves under Mahmoud Parvez's huge *shamianas*, erected on lawns illuminated with strings of multi-coloured bulbs and floodlighting. The Managing Director of the power station had just one worry: that one of the power cuts to which India was accustomed might plunge the festivities into darkness. To cover that eventuality, he had a powerful emergency generator set up behind one of the *shamianas*.

❉

A cool, bright winter's night had just fallen upon the City of the Begums. While preparations were going on in the Railway Colony and all the other places where weddings were due to take place, the married women in Orya Bustee had just finished dressing Padmini in her ceremonial clothes. Her father appeared at the entrance to the hut.

'Sister Felicity, look how beautiful my daughter is,' whispered Ratna Nadar proudly to the nun who had come to be with Padmini during the last moments of her adolescence.

'Oh yes, your daughter's very beautiful!' replied the Scotswoman, 'because God's loving hand created her.'

In a scarlet sari dotted in gold thread, her face concealed behind a muslin veil, her bare feet painted red, her toes, ankles and wrists glittering with jewels from the dowry brought by her future husband's emissaries, Padmini, escorted by her mother, was preparing to take her place on the rice straw mat placed in the centre of the *mandap*. It was there, beside the sacred fire burning in a small brazier, that she would await the arrival of the man whom destiny had given her as a husband.

Eyes shining with happiness, lips parted in a gratified smile, Ratna Nadar could not take his eyes off his child. It was the most beautiful sight of his life, a fairy tale scene, obliterating at a single stroke so many nightmare images: Padmini crying of hunger and cold on Bhopal station platform, foraging with her little hands through the piles of rubbish in between the rails, begging a few scraps of coal from the engine drivers . . . For this child of poverty-stricken parents there had been no play or schooling, only the supervision of her brothers, the drudgery of carrying water, doing laundry and household chores. It had been a life of slavery relieved only by her meeting with Sister Felicity. Today, dressed like a princess, Padmini would make the most of her happiness, her triumph, her revenge on a cursed karma.

A piercing cry, then the sound of moaning suddenly rent the night. A neighbour came running: 'Come quickly! Boda's having her baby.' Without a thought for her wedding clothes, Padmini dragged Sister Felicity to the hut where the wife of the dairyman Bablubhai was writhing in pain. Old Prema Bai was already there. Padmini held a candle over the thin, agonized face of the woman in labour. She was soaked in blood. Sister Felicity could see the child's skull showing between Boda's thighs. The young woman could not manage to expel it.

'Push,' urged the nun, 'push as hard as you can.'

Boda made such an effort that the tears poured down her cheeks.

'No, not like that, little sister! Push downwards. First try and breathe deeply, then push as you force the air out of your lungs. Quickly!'

Padmini lit a second candle to shed more light on Boda's lower belly.

'For the love of God, push harder!' begged the nun.

The dairyman's wife bore down with all her strength. Sister Felicity, who had assisted with dozens of births amongst the destitute, knew that this was their last chance of bringing the child into the world.

'Stand opposite me,' she ordered the old midwife who seemed overwhelmed by the situation. 'While I try and straighten the baby, you massage her stomach from top to bottom.'

As soon as the old woman started rubbing, the nun gently slid her hand behind the nape of the infant's neck. Boda let out a wail.

'Breathe deeply,' ordered the nun, 'and push regularly, without jerking.'

All the young woman's muscles tautened. With her head thrust back and her teeth clenched, she made a desperate effort.

The nun would never be able to explain what happened next. Her hand had just reached the baby's shoulders when two rats fell off the roof and passed in front of her eyes before landing on the stomach of the woman in labour. Taken by surprise, she withdrew her hand. Was it the suddenness of her movement or the shock occasioned by the creatures' fall? One thing was sure: all at once the child emerged.

Prema Bai cut the cord with her knife and tied it off with a strand of jute. The newborn baby was a fine boy who must have weighed nearly six pounds. Padmini watched as he filled his lungs and opened his mouth and let out a cry that was greeted with a tremendous echo of joy inside the hut and out into the alleyway.

'Sister, you've given me a son!' The dairyman Bablubhai was overjoyed. He brought a bowl full of rice, which he held out to Sister Felicity. 'Put this rice next to my boy, that the goddess may grant him a long and prosperous life.'

Then he called for an oil lamp. Traditionally, its wick must burn without interruption until the next day. If it went out, it would be a sign that the child born on this Sunday blessed by the stars, would not live.

❋

The magic moment in Padmini's life had at last arrived. A brass band burst into play, accompanied by singing. Preceded by a troupe of dancers outrageously made up with kohl, the groom's procession made its entry onto the parade ground outside the tea house. When she saw the boy astride his white mount, Sister Felicity thought she was witnessing 'the appearance of a prince from some Eastern legend'. His cardboard crown sparkling with spangles, his brocade tunic over white silk jodhpurs, and his mules encrusted with glass beads thrust into shining stirrups, the former little ragpicker and train scavenger looked like one of those Indian rulers popularized in engravings. Before climbing onto the *mandap* where his bride awaited him beside the sacrificial fire, Dilip had to submit to the ritual of purdah, wearing a veil so that his betrothed's eyes might not behold him before the appointed moment. He was then invited by the master of the ceremonies to sit down beside Padmini. Belram Mukkadam had donned an elegant brand-new white *kurta* for the occasion. Before the sun went down he had wanted secretly to conduct his own private celebration. He had tied his bull Nandi, bought with Carbide's compensation money, to the trunk of an acacia tree, painted his horns red and decorated his forehead with a trident, the emblem of the god Shiva. With this tribute, Mukkadam sought to invoke the sacred animal's blessing on the union of Dilip and Padmini.

'In the kingdom of heaven, theirs will be the most beautiful faces,' thought Sister Felicity as she looked round the men, women and children, in their festival clothes, encircling the bride and groom. With her bowed head partially concealed by her veil, Padmini seemed deep in meditation. This was the moment in which the nun chose to do something close to her heart. Sister Felicity got up and walked to Padmini.

'This little gold cross was given to me by my mother when I consecrated my life to God,' she said, fastening the chain round the girl's neck. 'It has protected me. Now I'm giving it to you so that it may protect you.'

'Thank you, Sister, I shall wear it always in remembrance

of you,' whispered Padmini, her eyes bright with emotion.

Then began the long ritual of an Adivasi marriage, punctuated with mantras in Sanskrit, the language of the sacred texts which Mukkadam had learned off by heart for the occasion, though neither he nor anyone else there understood them. He began by asking the couple to plunge their right hands into a baked clay jar filled with a sandalwood and tuber paste. In it two rings were hidden. The first to find a ring had the right to extract a forfeit from the other. After this preamble came the *panigrahan*. For the Adivasis, as for Hindus, this was an essential part of the marriage rite. The officiant took from the pocket of his *punjabi* a small piece of mauve cord and, taking hold of the couple's right hands, tied them together as he repeated their names aloud. The culminating moment had arrived. The band and the congregation fell silent. Now Mukkadam invited the married couple to officially make each other's acquaintance. Slowly, timidly, each parted the other's veil with their free hand. Dilip's delighted face appeared before Padmini's big, slanting eyes. Her heart was pounding. Her mother, father and brother watched her with barely contained emotion. Dalima, for her part, could no longer hold back tears. Already Mukkadam was asking the pair to complete the last part of the ceremony: with their right hands still bound together by the piece of cord, husband and wife walked seven times round the sacrificial fire.

It was ten o'clock at night and the celebrations were only just beginning. Helped by a group of women, Dalima started laying out on the sisal mats that had been rolled out near the tea house, plates made out of banana leaves. All the guests from Orya Bustee would soon sample the wedding banquet, looking out over the strange towers and pipework of the Carbide factory, lit up with strings of lights like an ocean liner.

While other marriage ceremonies were taking place in all four corners of the city, several hundred people were preparing to pay tribute to the Goddess of Poetry in Spices Square.

The organizers of that Sunday evening's *mushaira* had wanted to give their programme particular lustre by inviting one of the most famous Urdu poets. Jigar Akbar Khan was a legend in his own lifetime. In Bhopal he had such a following that a taxi driver had once abducted him to force him, at gun point, to give him a private recital. Jigar could declaim more than fifty *ghazals* in a single evening. Whenever he appeared, his audience went into a frenzy. His sublime incantations, his sonorous voice, sometimes caressing, sometimes imploring, were magical. It was common knowledge that the elderly bearded poet, enveloped in his shawl, was the most hopeless drunkard, but what did that matter? Bhopal was indebted to him for too many nights of exaltation not to forgive him. It was said that one of his disciples had actually left his wife on their wedding night to accompany the master back to the railway station and put him on his train. Just as the train was pulling away, the waggish Jigar had grabbed hold of his admirer and prevented him from jumping back onto the platform. The newly wedded husband had not returned to Bhopal for a year, a year spent following his idol from festivals to *mushairas* across India.

With their hands outstretched towards the narrator in a gesture of offering, eyes closed upon some vision of ecstasy, gently shaking their heads as a mark of their approval, the audience greeted each verse with an enthusiastic 'Vah'[1]. The

---

[1] Marvellous!

chill of a slight breeze blowing from the north nipped at their flesh, but exaltation warmed bodies as well as souls.

Was it a premonition? The elderly poet with the white beard began his recital with an evocation of the suddenness of death:

> *Death which appears*
> *Like a silent dragonfly*
> *Like the dew on the mountain*
> *Like the foam on the river*
> *Like the bubble on the spring . . .*

PART THREE

# THREE SARCOPHAGI UNDER THE MOON

# 36

## *Three sarcophagi under the moon*

In their reinforced concrete tomb, the three tanks, two yards high and thirteen long, looked like enormous sarcophagi left there by some pharaoh. They lay, half buried, side by side, at the foot of the metal structures on view to Dilip and Padmini's wedding guests. They had no names on them, only numbers: E610, E611, E619. These tanks were masterpieces of the most advanced metallurgy. No acid, liquid or corrosive gas could eat into their shells which were made out of SS14 stainless steel. At least, that was the theory: methyl isocyanate had not revealed all of its secrets. A complex network of pipes, stopcocks and valves linked the tanks to each other and to the reactors that produced the MIC and Sevin. To prevent any accidental leakage of their contents into the atmosphere, each tank was connected to three specific safety systems. The first was a network of fine piping which covered the tanks. The freon gas flowing through it kept the MIC constantly refrigerated to a temperature close to 0° celsius. The second was a monumental cylindrical tank called a 'decontamination tower'. It contained caustic soda to absorb and neutralize any escaping gas. The third was a 120 foot flare. Its role was to burn off any effluents that had escaped the barrage of caustic soda.

That 2 December 1984, there were forty tons of methyl isocyanate in the tanks – a 'real atomic bomb right in the middle of the plant' according to the German chemist from Bayer once questioned by Eduardo Muñoz – and not one of the three safety systems was operational. The refrigeration had been stopped for a month and a half and the MIC was being kept at the ambient temperature, about 20° celsius in a winter month. The alarm that was supposed to go off in case of any abnormal rise in temperature in the tanks had been disconnected. As for the decontamination tower and the flare to incinerate the gases, several of their components had been dismantled the preceding week for maintenance.

No mention was made of it in the technical handbooks, but there was a fourth safety device. No corrosion or cutbacks could put it out of action because the only power source this funnel-shaped piece of material needed was the God Aeolus's breath. The windsock fluttering over the factory supplied an essential piece of information: the wind direction. Lit up at nightfall, it was visible from all workstations. The occupants of the surrounding neighbourhoods, however, could not see it. No one had thought to put another over their bustees.

&#10034;

There were further grounds for concern. With forty tons of MIC inside, tank 610 was almost full, and that was in absolute violation of the safety regulations. The tanks were never meant to be filled to more than half their capacity, just in case a solvent had to be injected to stop a chemical reaction. Tank 611, next to it, contained twenty tons of MIC. As for the third tank, 619, which was supposed to remain empty to act as an emergency tank in case of an accident in the other two, it contained one ton of MIC.

Since 26 October, the day on which the factory stopped production, the contents of these tanks had not been analysed. That was another serious breach of regulations. Methyl isocyanate is not an inert substance. Because it is made up of multiple gases, it has a life of its own and is constantly changing and reacting. Was the MIC inside the three tanks still the 'pure, clear mineral water' Shekil Qureshi, Pareek's young assistant, had once admired? Or had it been polluted by impurities likely to cause a reaction? Broken down by heat, the MIC could then emit all kinds of gases, including the deadly hydrocyanide acid. In the event of a leak, these gases with different densities would form toxic clouds which would spread at different speeds and on several levels, saturating a vast area in one fell swoop.

The level of deterioration the plant had reached meant that the worst could be anticipated. Moreover, there were indications that strange things were going on that night, in tank 610 as well as in the apparatus next to it. Twice in succession, on 30 November and the following 1 December, operators had tried to transfer some of the forty-two tons of MIC to the unit manufacturing Sevin. In an operation of this kind, the contents of the tank had first to be pressurized by introducing nitrogen, a routine process in a properly maintained factory. But the 'beautiful plant' was no longer in very good shape. Because of a defective valve, the nitrogen escaped as fast as it was put in. The valve was not replaced, and the forty-two tons of MIC were left in a tank that had not been properly pressurized. This meant that any contaminating substance from outside could get into the tank without meeting any resistance, and thus trigger an uncontrollable chemical reaction.

❈

Rehman Khan was a young, 29-year-old Muslim who seldom parted from his embroidered skullcap even when wearing his safety helmet. Originally from Bombay, he had moved to Bhopal to get married. His wife worked as a seamstress in the workshop that made the workmen's boiler suits. It was thanks to her that, after a brief training period, he had joined the MIC production unit as an operator. He had been working there for four months and earned 1400 rupees, a comfortable salary given his lack of experience and qualifications. Like most of the 120 unskilled workers on the site that evening, he had practically nothing to do. The factory had been stopped. Khan was part of the second shift, and was on duty until 23.00 hours. A passionate lover of poetry, as soon as his shift was over he intended going to the grand *mushaira* being held in Spices Square for the festival of Ishtema. To kill time that dreary winter's evening, he was playing cards with some of his comrades in the canteen when a telephone call summoned him urgently to see the duty supervisor. Gauri Shankar, a tall, bald Bengali, seemed extremely irritated.

'That lazy maintenance team hasn't even managed to flush out the pipes!' he grumbled.

Shankar was referring to the pipework carrying to the tanks the liquid MIC produced by the plant's reactors. Because of its highly corrosive nature, methyl isocyanate attacks pipes, leaving scoria deposits on their lining. High pressure jets of water had constantly to be sent into the piping to get rid of these impurities, not just because they would eventually block the flow, but above all because they could get into the storage tanks, contaminate the MIC and set off a reaction with uncontrollable consequences.

Shankar brandished the logbook containing remarks related to the running of the production unit.

'Here are the instructions left by A. V. Venugopal,' he

explained. 'The production supervisor is asking us to flush the pipes.'

Khan knitted his thick eyebrows.

'Is it absolutely necessary to do it this evening? The plant's stopped. I would have thought it could wait till tomorrow. Don't you think?'

Shankar shrugged his shoulders. He had no idea. In truth neither he nor his supervisor, Venugopal, were familiar with the factory's very complex maintenance procedures. They had both only just arrived there, one from Calcutta, the other from Madras. They knew virtually nothing about MIC or phosgene apart from their very distinctive smells. The only industry they were familiar with was the one that produced Carbide's batteries.

In his note, the supervisor had given succinct instructions as to how the requisite washing operation should be carried out. He stipulated that it should begin with the cleaning of the four filters and the circuit valves. He went on to supply a list of stopcocks to be turned off to prevent the rinsing water from entering the tanks containing the MIC. But he had forgotten to recommend one crucial precaution: the placing of solid metal discs at each end of the piping connected to the tanks. Two segments of the pipework had only to be disconnected and the discs slid into housings provided for the purpose, then the whole thing bolted up again. The process took a little less than an hour. Only the presence of these 'slipbinds' as the engineers called them in their jargon, could guarantee that the tanks were hermetically sealed. The valves and stopcocks under attack from corrosion could not, alone, ensure their insulation.

Rehman Khan first set about closing the main stopcock. It was a complicated process because the stopcock was located three yards off the ground, at the centre of a tangle of pipes that were difficult to get to. Bracing himself against two girders, he put all his weight on the handle that closed the stopcock, yet he still could not be sure that he had managed to seal it completely, so rusted and corroded were the metal parts. After that, he climbed back down to turn off the other stopcock and start flushing. He had only then to connect a hosepipe to one of the draincocks on the pipework and turn on the tap. For a few seconds he listened to the water rushing vigorously into the pipes and noted the time in the logbook: it was 20.30 hours.

The young operator very swiftly realized that something unusual was going on: the injected water was not, as it should have been, coming out of the four draincocks provided for the purpose. Khan tapped them lightly with a hammer and discovered that the filters in two of them were blocked with metal debris. He immediately cut off the water supply and alerted his supervisor by telephone. The latter did not arrive for quite a while, and when he did, his lack of experience meant he was not much help.

He simply instructed Khan to 'clean the filters on the evacuation draincocks well, and turn the water back on. With the pressure of the flow, they'll let the water out eventually.'

The young Muslim agreed with some reservations.

'But if the water doesn't come out through the draincocks, it'll go somewhere else,' he suggested.

The supervisor failed to grasp the vital implications of this remark.

'We'll just have to see!' he replied, irritated at having been disturbed for something so trivial.

As soon as his boss had gone, Khan set about cleaning the filters, then turned the wash tap back on. Shankar was right: the water flowed out normally through the first two draincocks and, after a moment, through the third one too. But the fourth seemed to be permanently blocked. Khan was not unduly worried. As his boss had said, the system would

266

eventually clear itself. He went on flushing the pipes, using all the pressure in his hose. Several hundred gallons poured into the pipes. Two hours later, at 22.30, half an hour before the changeover of shifts, he knocked on the door to his boss's cabin.

'What shall I do?' he asked. 'Shall I keep the water running, or should I turn it off?'

Shankar looked doubtful. He rubbed his chin.

'Keep it running,' he eventually said. 'The insides of those bloody pipes are supposed to be completely spotless. The night shift will turn the tap off.'

At these words, Rehman Khan took a pencil and wrote a brief report of the operation in progress in the logbook.

'Good night, Sir. See you tomorrow!' he called out, in a hurry to have his shower and dress for the evening's big event, the *mushaira* in Spices Square.

It was eleven o'clock at night. Spices Square was humming with poetry lovers impatient to hear their favourite poets. On the other side of the city, the reception rooms and lawns of the Arera Club were teeming with guests, as were the sumptuously decorated tents set up for the marriages in the rich districts of New Bhopal and Shamla Hill. On the Kali Grounds, strings of bulbs lit up Dilip and Padmini's wedding celebrations. The whole of Bhopal was giving itself up to rejoicing on that night blessed by the stars. It was in the Railway Colony that the festivities taking place beneath a shower of fireworks were most splendid. The one thousand guests at the wedding of Rinou Diwedi, younger daughter of the chief controller of the Bhopal railway, to the son of a Vidisha merchant, watched with wonder the ceremonial procession of the Barat. Perched on a white mare covered with a velvet cloth embroidered with gold, wearing a spangled turban on his head, young Rajiv caracoled towards his fiancée, who was waiting for him under Parvez's most beautiful *shamiana*. Before he straddled his mount, his father had imprinted on his forehead the two red and black spots that would banish the evil eye for ever and guarantee him a propitious future. Then Rajiv had been given a coconut with red stripes scratched onto it, a traditional token of good luck. In front of the white mare walked a woman, taking tiny steps: his mother. She was dressed in the double silk and gold sari she kept for special occasions. Fervently she strewed the ground with handfuls of salt, to eliminate from her son's path all life's possible pitfalls.

# 37

## 'What if the stars were to go on strike?'

Twenty-three hundred hours. It was time for a change of watch on the bridge of the vessel Rehman Khan and his comrades from the previous shift had just left. The man who took over command of the control room was a Bengali Hindu. He was twenty-six years old and his name was Suman Dey. A science graduate of the University of California, he was both competent and respected. The seventy-five dials lit up in front of him made up the factory's control panel. Every needle, every luminous indicator supplied information, showed the state of activity in each section, signalled an eventual anomaly. Temperatures, pressures, levels, outputs – in his capacity as officer of the watch, Suman Dey was kept constantly up to date with the condition of the plant. At least that was the theory, but for some time now some of the apparatus had been breaking down. Dey was therefore obliged to go and get his information on site. He was not always able to. For several days now, because of a fault in the transmission circuit, there had been no temperature reading coming through from tank 610. To calm his own frustration, he meditated on the words of a large notice hanging on the wall above the dials: *Safety is everybody's business*. Nothing definite, however, made the young Bengali think that the safety of the factory was not assured.

There was not the least sign of disquiet on the faces of the six night shift operators. They settled in for the night around the brazier in the small room adjoining the control room known as the site canteen because those in it could be mobilized immediately in case of alert. The men on that night shift were a perfect reflection of India's enormous diversity. Next to the Muslim supervisor, Shekil Qureshi, the man who had escorted the MIC trucks, sat the Sikh, V. N. Singh, whose parents had been so thrilled to see him join Carbide. Third was a tall, 29-year-old Hindu with a melancholy face. Mohan Lal Varma was in dispute with the management who, for six months, had been refusing to give him his classification and salary as a sixth grade operator. Fourth was a Jain, originally from Bombay and as thin as a wire; finally, there were the son of a railway employee from Jabalpur and a former trader from Bihar.

Apart from Qureshi, Singh and Varma, who were to continue the cleaning operation that the previous shift had started, the men had nothing specific to do that night because their production units had been stopped. They chatted about the 'beautiful plant's' gloomy future, smoking *bidis*, chewing *pan* and drinking tea.

'Apparently the Sevin sales aren't going too well any more,' said the Jain from Bombay.

'They're going so badly that they've decided to dismantle the factory and send it in bits to some other country,' added the former merchant from Bihar, who had become a specialist in alpha naphtol.

'Which country?' the Jain asked anxiously.

'Venezuela!' asserted the Muslim from Jabalpur.

'Not Venezuela!' corrected Qureshi, who had his sources in the management offices: 'Brazil.'

'Meanwhile, we're the ones Carbide drops in the shit,' interjected Varma angrily. His struggles with his superiors had made him aggressive.

Qureshi tried to allay his colleagues' fears. They all liked this tall, slightly clumsy fellow, who was always ready to share his inexhaustible repertoire of *ghazals*. The nights did not seem quite so long when they could listen to him singing his poems. They had been pleasantly surprised to find him there that evening, because the roster had not shown him on duty until the next day. At the last minute, however, he had agreed to stand in for a colleague who had been invited to one of the weddings: a very noble gesture on his part on a night when there was a *mushaira*.

While still carrying on with the discussion, Qureshi cast an eye over the logbook, brought up to date by the previous shift. On the page for tank 610, for the pressure reading for 20.00 hours, he read: '2 p.s.i.g.'. He gave a smile of satisfaction. Two pounds per square inch of pressure! That meant that all was well inside the tank. Suddenly the Muslim's expression darkened, when he realized that this information was three hours old. Three hours!

'Before half the technicians were laid off, we used to take pressure and temperature readings every two hours. Now it's every . . .'

'Eight hours,' specified Suman Dey, who had just emerged from the control room.

At the beginning of that December, an atmosphere of extreme depression prevailed over the metal platforms of the Kali Grounds. Ever since the departure of the men who had given the factory its soul – Woomer, Dutta, Pareek, Ballal . . . morale had plummeted, discipline had lapsed and, worst

271

of all, the ideal of safety had gone out the window. It was rare now for those handling toxic substances to wear their helmets, goggles, masks, boots and gloves. It was even rarer for anyone to go spontaneously in the middle of the night to check the welding on the pipework. Eventually, and insidiously, the most dangerous of ideas had crept in, namely that nothing serious could happen in a factory when all the installations were turned off. As a result, people preferred card games in the site canteen to tours of inspection round the dormant volcano.

※

'Hey guys! Can you smell it? Hey, can you smell it?' Mohan Lal Varma had sprung to his feet. He sniffed noisily. 'Have a sniff, go on! I swear there's MIC in the air!'

This sudden excitement on the part of the usually passive young Hindu provoked much amusement all round.

'Sort your hooter out, idiot!' cried the Jain from Bombay. 'There can't be any smell of MIC in a factory that's stopped!'

'It's not MIC you can smell, it's Flytox!' interrupted the factory worker from Bihar. 'They sprayed a whole canister of it about before we got here!'

'That's why we haven't been bombarded with mosquitoes yet!' confirmed the Muslim from Jabalpur.

Everyone in Bhopal agreed: Flytox was a magic invention, a real godsend, the miracle insecticide that provided protection against the City of the Begums' worst scourge: its mosquitoes.

※

Amidst all the hullabaloo of the festivities taking place on the other side of the Kali Grounds, no one noticed a frail young girl dressed in a simple blouse and blue cotton skirt. She threaded her way through the assembly preparing to dine on the sisal mats. She approached several of the guests, apparently looking for someone.

'Do you know where Sister Felicity is?' she asked, clearly agitated.

Dalima, who had overheard the question, joined the stranger and scrutinized the faces by the light of the strings of bulbs. The banquet had begun. The men were on one side, the women on the other. Only the bride was missing from the feast. She had momentarily withdrawn to a neighbour's hut to open her wedding presents. In the end, Dalima spotted the missionary sitting amongst a group of women. She directed the young messenger over to her.

'Anita, what are you doing here?' the nun asked, surprised.

'Sister, you must come at once, there's been an accident at home.'

The Scotswoman led Anita to a motorized rickshaw parked outside the tea house.

'What is it?' she asked, concerned.

'The little one you have in your room . . .'

'Nadia?'

'Yes. She had a terrible fit. She started smashing everything up. She yelled far more loudly than last night, louder than any of the nights before the monsoon. She yelled like a madwoman. She called for you. Three of us tried to calm her down, restrain her, but . . .'

'But?'

'She got away from us. She threw herself out of the window.'

'Oh my God! . . . —' The nun felt her heart pound. For a few seconds she remained silent, then, slowly crossing herself, she said softly: 'Lord Jesus, receive your innocent child into your Paradise.'

'She's not dead, Sister!' Anita went on quickly. 'An ambulance has taken her to Hamidia Hospital.'

Fifteen minutes later, Anita and Sister Felicity ran through the emergency entrance to the building where the air was thick with the smell of disinfectant and ether. The floor was spotted with red stains left where people chewing betel had spat. The wards were almost empty. Sunday was not a day for accidents. Under the inscription DOCTORS ON DUTY, two doctors were settling down for a quiet night in their small office. Tall and lanky, with his thick black shock of hair carefully combed, the 35-year-old Hindu, Deepak Gandhe, and his young Muslim colleague Mohammed Sheikh had been students together at the Gandhi Medical College, the enormous building on the other side of the road. Since then they had been inseparable. One was a general practitioner; the other a surgeon. That was the usual combination for a tour of duty. The arrival of Sister Felicity and the young Indian girl caught them right in the middle of a game of dominoes. They stood up.

'Doctors, we've come about Nadia,' said Sister Felicity.

Dr Sheikh's face froze. He played nervously with his moustache. The two women prepared themselves for the worst. Dr Gandhe, however, gave the faintest of smiles.

'Little Nadia has undergone an operation,' he confided softly. 'For the moment she has survived her injuries. We hope to be able to save her. She's in intensive care.'

The Scotswoman's eyes filled with tears.

'May I see her?'

'Yes, Sister, you can even spend the night with her. You'll have the whole ward to yourself. There's no one else in intensive care this evening.'

❉

While Sister Felicity and young Anita began their night of prayer beside little Nadia's injured body, the 1000 guests at the wedding in the Railway Colony tucked into petits-fours, kebabs, prawns, diced chicken in ginger, and pieces of cheese wrapped in spinach, brought round by an army of turbaned servants. Despite the fact that his cardiologist had forbidden him alcohol because of his coronary problems, Harish Dhurve, the stationmaster, chanced his luck with the glasses of 'English liquor', whisky imported from Britain. Suddenly he found himself nose to nose with his doctor.

'Indulge me, doctor, this evening is exceptional, a night blessed by the stars!' he apologized.

Dr Sarkar was the official doctor for the residents of the Railway Colony and the station staff. His Bengali sense of humour meant that he was never at a loss for repartee. Looking pointedly at his patient's glass, he asked:

'And what if the stars decided to go on strike?'

This reply brought a slightly forced smile to the stationmaster's face. More than anyone else in Bhopal that night he needed the blessing of the stars. Like most of the other railway employees invited to the festivities, he would have to slip away a little before midnight to attend to his duties at the station. In fact that night was expected to be extremely busy because of the pilgrims arriving to celebrate Ishtema. Dhurve had had all staff requisitioned, including the 101 coolies. His station was one of the country's principal railway junctions. He had promised himself that he would control the excess traffic with punctuality and suppleness, and provide the thousands of visitors with a welcome befitting Bhopali hospitality.

Midnight. In the factory, unbeknown to anyone, a bomb had just been primed. After the night shift operators had tried vainly to drain the system of the rinsing water that had been injected into it for the last three hours, it had started to blow back into tank 610. It went rushing in, carrying with it metal debris, sodium chloride crystals and all the other impurities it had dislodged from the lining of the pipes. This massive influx of contaminants promptly set off the exothermic reaction the chemists had so dreaded. In a matter of minutes, the forty-two tons of methyl isocyanate disintegrated in an explosion of heat, which would very quickly transform the liquid into a hurricane of gas.

When their eyes began to smart, the six men sitting in the site canteen less than forty yards from the tanks, finally conceded that their colleague Varma was right. It was not the smell of Flytox he had detected, but indeed the characteristic boiled cabbage of methyl isocyanate. They still did not know, however, what was going on in tank 610.

Qureshi turned to V. N. Singh and Varma.

'Guys, you'd better go and do a tour round the rinsing area,' he suggested. 'You never know.'

The two technicians picked up their torches, put on their helmets and stood up.

'Don't forget your masks!' recommended the Muslim supervisor.

'It's not worth it! It's not the first time this factory's smelt of MIC,' replied V. N. Singh. 'Have the tea ready for us in a minute!'

'Of course!' called out Qureshi.

'And if you're not back in time we'll send out a search party with a bottle of oxygen!' joked the Jain from Bombay, provoking general laughter.

In a few minutes, the two men reached the pipework being cleaned. The smell was getting stronger and stronger. They listened to the rushing of the water still circulating at full force through the piping, and directed the beams of their torches onto the network of pipes. They scrutinized every stopcock, valve and flange. All of a sudden Singh noticed round a draincock some eight yards off the ground a bubble of brownish water surmounted by a small cloud. He drew his workmate's attention to it.

'There's some gas escaping up there!'

Varma pointed the beam of his torch at the cloud.

'You're right. And it's not Flytox!'

The two men ran back to the control room.

'Shekil! There's a pipe pissing MIC!' announced Singh. 'You should come and take a look.'

Qureshi looked at his colleague in disbelief.

'Stop fooling about!' he protested. Then emphasizing each word, he insisted: 'Get it into your heads once and for all that there can't be a leak in a factory where production has been stopped. Any idiot knows that.'

'But it really is pissing out, and it smells very strong!' insisted Singh, rubbing his eyes.

Qureshi shrugged his shoulders.

'Perhaps it's a drop of residual MIC escaping from the drainage cocks with the rinsing water,' he conceded. 'All we have to do is turn the water tap off. We'll see if we can still smell it after that.' With these words he looked at his watch and added: 'For now, guys, it's midnight and time for tea!'

The sacrosanct tea break! Thirty-seven years after their colonizers had left, no Indian, not even the six Carbide men perched atop an erupting volcano, would forego a ritual that had entered their culture as surely as the passion for cricket. Qureshi led the team to a building a hundred or so yards away, which housed the staff cafeteria. Shortly after midnight, a young Nepalese lad with small, laughing eyes made

277

his appearance. He was the tea boy. In his basket he carried a kettle full of scalding-hot, milky tea, some glasses and a plateful of chocolate biscuits.

Qureshi and his workmates settled down comfortably to sip the delicious brew steeped in the rich perfume of the distant hills of Assam. Suddenly, a worried face appeared in the doorway. It was Suman Dey, the duty head of the control room.

'Shekil,' he called out to the Muslim supervisor, 'the pressure needle for tank 610 has shot up from two to thirty p.s.i.g.!'

Qureshi shrugged his shoulders, then gave his colleague a smile: 'Suman, you're getting in a sweat about nothing! It is your dial that's up the creek!'

# 38

## Geysers of death

Stations along the world's second largest rail network knew nothing about closing for the night. In Bhopal, platform no 1, the same platform that, a hundred years previously, had greeted the Kingdom of the Begums' first train with a double line of mounted lancers and turbaned sepoys, was seething with activity. That night it was swarming with hundreds of passengers waiting for the Gorakhpur Express. As a precaution against thieves, many of them had chained their luggage to their ankles. Tormented by mosquitoes, hordes of children were running about in all directions, playing hide and seek amongst the suitcases and squabbling amongst themselves. Dozens of street vendors, porters in red tunics, lepers shaking their bowls and ringing their bells, beggars, and policemen in blue caps were wandering amongst the travellers and their luggage in an acrid atmosphere of *bidi* smoke, betel, and incense sticks.

Midnight was the time for the shifts to change. V. K. Sherma, the deputy stationmaster, and Madan Lal Paridar, his assistant, together with their young aid, the traffic regulator Rahman Patel, had just settled themselves in front of the control board in their office at the end of the platform.

279

With its Victorian gothic architecture, it looked like a Sussex cottage. The room was equipped with two powerful air-conditioning units, which in the summer made it possible to forget the heat and pollution outside. Now, however, it was winter and the machines were switched off. Cool air from outside came in through the wide open doors and windows. Inside, the office was equipped with a long board on which moveable pegs of different colours and lights marked the location of trains on their way to Bhopal as well as the position of the signals and points. On the table in the middle of the room stood several telephones, one of which was an old-fashioned crank phone that they used to call other stations to check what time the trains went through. Because of thick fog over part of Madhya Pradesh that night, and the unusual amount of activity, most of the trains due to arrive before midnight showed significant delays. None of them was expected before two or three in the morning. Such was the case with the Gorakhpur Express, in which Sajda Bano, the widow of Mohammed Ashraf, Carbide's first gas victim, was travelling. Together with her two sons, 3-year-old Soeb and 5-year-old Arshad, she had meant to catch the train on the previous day. At the last minute, however, a neighbouring Hindu woman had begged the young Muslim not to travel on Saturday, because it was considered by followers of her religion to be the most ill-omened day of the week.

In Bhopal station, to the incessant sound of jangling telephones, the staff prepared for one of those long waits, to which Indian railway employees and their twelve million daily passengers are well accustomed. Suddenly, the deputy stationmaster picked up one of the phones.

'I'm calling the boss,' he informed his colleagues. 'There's no need for him to come for a good two hours yet.'

'You're right,' approved his assistant, 'that way he'll be

able to get a few more small glasses of English liquor down him!'

The three men laughed. They were well aware of Harish Dhurve's weakness for alcohol. It was at this point that a coolie in a red tunic appeared at the door.

'Boss! Boss! Quickly, come and see! Arjuna and his chariot are here with presents for you.'

The porter Satish Lal was in a state of extreme excitement. His reference to the mythological Pandava prince and his celestial chariot was not wholly inaccurate. Ratna Nadar, whom he had introduced to his team of porters two years previously, had just arrived, pushing a rickshaw full of small cardboard boxes.

'There are a hundred and five, one for each coolie, four others for the bosses and the last is for good old Gautam behind his ticket office window,' announced Padmini's father.

Each box contained a hard-boiled egg, a kebab on a stick, a small bowl of rice with dal on it, a vegetable samosa, a chapatti and two balls of *rossigola*[1]. In India every feast is shared. Ratna Nadar had been eager for his workmates and superiors to have their share of the banquet held that night to mark this most important event in his life: his daughter's marriage. Towards eleven o'clock, he had slipped away from the festivities to change from his ceremonial clothes into his working ones. That night, he too had been requisitioned to assist the expected passengers.

'Ratna Nadar *Ki Jai*! Ratna Nadar *Zindabad*!' A rousing ovation in Hindi and Urdu acclaimed the father of the bride and his cartload of food.

'Thank you, friends! Thank you, my friends!' he repeated over and over again as he handed out his little boxes.

Drawn by the unusual cluster of red tunics right in the

---

[1] Very sweet confection made out of pastry steeped in syrup.

middle of the platform, some of the passengers gathered round. Ratna Nadar cast an emotional eye over the dilapidated façades of the vast station where once he had disembarked with all his family, driven from his village by marauding insects; the same station which today incorporated all his hopes. Thanks to it, to all of its passengers' mountains of bags and packages, and to the heavy crates in its cargo bays, he was going to be able to pay back the 12,000 rupees borrowed from Pulpul Singh for his daughter's wedding. Every train would bring him nearer to that blessed day when he would be able to recover the property deeds for his hut that he had pawned to the money lender.

⁂

Less than half a mile away as the crow flies, the curtain was rising on the tragedy of which Rajkumar Keswani had forewarned the people of Bhopal. The supervisor Shekil Qureshi showed no signs of hurrying his cup of tea. In his view the man in charge of the control room was over-reacting. He knew that thirty pounds of pressure per square inch were not really grounds for alarm. The South Charleston engineers had designed the MIC tanks with special steel and walls thick enough to resist pressures five or six times greater. But the needle on the dial in the control room had now leapt up again, to 55 p.s.i.g, which was at the upper end of the scale on the dial. More importantly, it was twice the limit the engineers referred to as the 'permitted maximum working pressure'. Was the instrument malfunctioning as Qureshi supposed, or was the pressure it was showing real? For Suman Dey, there was only one way to find out: by going into the zone where the three tanks were, to look at the pressure gauge directly attached to tank 610. If it confirmed the figures on the control room dial, then something out of the ordinary was going on.

'Let's go, Chandra!' said Dey to one of the operators on duty.

'Are we taking masks?'

'Not half! Masks and bottles!' insisted Dey, who had a visceral fear of chemical substances.

Each bottle was guaranteed to last half an hour. When it was down to just five minutes' worth of oxygen, an alarm went off.

It took less than three minutes for the two men to get to tank 610 and establish that the needle on the pressure gauge was also indicating 55 p.s.i.g. Dey climbed onto the concrete sarcophagus in which the tank was imbedded, knelt down on the top, took off his glove and palpated the metal casing meticulously.

'There's a hell of a lot of movement going on in there!' he shouted through his mask.

The stirring he had felt was the now gaseous methyl isocyanate sweeping into the pipes leading to the decontamination tower. That was where it was supposed to go in such circumstances. But, that night, the stopcocks controlling access to the safety device were turned off because the factory was not in service. Under pressure that was mounting by the minute, the column of gas was popping bolts like champagne corks. Some of the gas then escaped, giving rise to the sort of small brownish cloud that operators V. N. Singh and Varma had spotted before their tea break. Both had returned hastily to the zone where the pipes were being cleaned, this time equipped with masks and oxygen bottles. The first thing they did was to close off the water tap, turned on four hours earlier by their colleague Rehman Khan. Even with their masks on, they could smell powerful gas emissions.

283

'It stinks of MIC and phosgene too,' grunted V. N. Singh, who had recognized the characteristic smell of freshly mown grass.

'And of MMA!' added Varma, picking up the suffocating smell of monomythylamine ammonia.

A hissing noise like that of a jet of steam was suddenly heard overhead. Instantly they looked up at the network of pipes. A geyser had just burst from the spot where they had first detected the gas leak. Despite his terror V. N. Singh managed to keep a cool head. There was only one thing to do in such circumstances. He had done it before at the time of the great fire in the alpha naphtol unit. He hurled himself at the nearest alarm point, broke the glass, and pressed on the button that set off the general alarm siren.

*

The siren's howl wrested Shekil Qureshi from his cup of tea. He ran out of the cafeteria and rushed to the control room where he met V. N. Singh, who had just come back up from the pipe cleaning zone. Singh took off his mask. He was a ghastly colour.

'The worst has happened. There's nothing we can do,' he muttered, shaking his head, overwhelmed.

Qureshi protested fiercely:

'It must be possible to contain this bloody reaction. I'm going quickly to see what's going on.'

Singh called after his disappearing figure:

'Your mask!'

'Can't give orders with that thing over my face!' replied the Muslim, who was already scrambling down the stairway.

When he got to the erupting geyser, he stopped dead in his tracks. He could not believe his eyes. 'It's not true! . . .' he murmured. There he was, the man who had been so convinced that no accident could happen in a factory that was not running, witnessing precisely the catastrophe which all Carbide's manuals, all its safety exercises and all its security campaigns had persistently warned against: a terrifying, uncontrollable, cataclysmic exothermic reaction of methyl isocyanate. A massive reaction of a whole tank full, not just of a few drops left in a pipe. How had such an accident come about, despite all the safety regulations? Qureshi beat a retreat and made for the zone where the tanks were. He had an idea. Even if it was too late to stem the eruption of tank 610, at least the contamination could be prevented from reaching the twenty tons stored in tank 611. His eyes were beginning to burn painfully. He was having progressively more difficulty breathing. In a blur he saw Suman Dey and his companion coming down from the sarcophagus onto which they had courageously climbed to check the pressure indicator. The tank and its concrete casing were trembling, cracking and creaking as if shaken by an earthquake. The voice of the Muslim supervisor was faintly audible through the chaos.

'We must isolate 610! We must isolate 610!' He shouted himself out of breath.

Suman Dey did not agree. By turning off the valves and stopcocks connecting the reacting tank to its neighbour, they would risk increasing the pressure and possibly set off an explosion. But Qureshi had faith in the tank's capacity to resist anything. How could this technological masterpiece that he had once witnessed arrive from Bombay, this precious jewel, all the connections to which he had lovingly maintained, repaired and nurtured, possibly disintegrate like some common petrol tank? Dragging his two companions with him, he threw himself at the pipework. The ground was cracking beneath their feet. There was a noise as if the end of the world were coming. In ten minutes, they managed to shut off all communication between the two tanks. The twenty tons in 611 would not be caught up in the gaseous apocalypse.

They immediately retreated at a run. Before disappearing into the stairway leading to the control room, they turned round. Tank 610's concrete carapace had just shattered, releasing an enormous steel tank which emerged from its sarcophagus like a rocket, stood vertically, toppled, fell and stood up again before tumbling heavily onto the concrete and metal debris. But it had not burst. From a ruptured pipe at ground level a second geyser then erupted, more powerful and even fiercer than the first.

Before entering the control room Qureshi glanced at the windsock flying from the top of its mast. He grimaced. Filled by an unremitting wind, the white material cone pointed clearly south, towards the neighbourhoods of the Kali Grounds, the station and the old part of the city. That night, however, true Carbider that he was, he felt most responsible for his men and their safety. He turned to the chief of the watch:

'Suman, turn your siren off and yell into the loudspeakers. Get everyone to assemble in the formulation zone on the north side, except the operators in our unit who should remain available with their masks. We may need them later.'

# 39

## *Lungs bursting in the heart of the night*

For the supervisor Shekil Qureshi, the young Muslim who at his wedding had thought that the finest clothing he could ever wear would be 'the linen boiler suit with the blue and white logo', all was not yet lost. He wanted to attempt the impossible.

'Suman! Try and get the decontamination tower up and running,' he ordered the man in charge of the control room. 'You never know, perhaps the maintenance team has finished its repairs.'

Suman Dey tried the control lever, but there was no reaction on the dial on the control panel. The indicator did not light up and the pressure needle remained at zero.

The telephone rang. Qureshi picked it up. It was S. P. Chowdhary, the production manager, calling from his villa in Arera Colony on the other side of Bhopal. He had just been woken up by one of the night shift operators.

'I'll be there as quickly as I can!' he shouted into the phone. 'In the meantime try and get the flare going!'

Qureshi could not believe his ears. What? The man in charge of production at the factory was not aware that the emergency flare was undergoing repairs?

'The flare?' he queried indignantly, 'but there are five or six yards of pipe missing from it! They were rotten.'

'Replace them!' insisted the Production Manager.

Qureshi held the telephone receiver outside the window.

'Do you hear that? That's gas pouring out. Even if we were to manage to replace the pipes, we'd have to be out of our minds to light the flare. We'd all be blown up and the factory and the entire city with us!'

Furious, Qureshi hung up, but he still refused to admit defeat.

'Get me the fire brigade,' he asked Suman Dey.

Qureshi begged Carbide's fire brigade chief to send men as fast as possible to douse the geyser spurting out from under the decontamination tower. He knew that water, which could cause the methyl isocyanate to explode in an enclosed environment, could also neutralize it in the open air – a chemical contradiction that had induced the three American engineers who came to inspect the factory in 1982 to call for the installation of an automatic sprinkler system in the sensitive MIC production zone. Their request had been ignored, and as a result, men would risk their lives trying to act as human sprinklers.

In less than five minutes, the firemen were on the scene. Almost immediately, their chief's voice came over the radio speaker.

'Impossible to reach the leak! Our hose jets won't go that high!'

This time Qureshi realized that there was nothing more to be done.

'Give the order for everyone to evacuate, directly to the north,' he ordered Suman Dey, 'and let's get out of here fast!'

The proud Muslim rushed to the cloakroom to pick up his mask. But his locker was empty and his mask had gone. He had to escape with his face exposed. With his eyes burning, his throat on fire and gasping for breath, he ran like a madman. He thought of his wife and children. 'I was so afraid of dying, I felt capable of anything,' he said later. What he did was scale the two-yard high perimeter wall of the factory and the coils of barbed wire on top and drop down on the other side. In his fall, he tore his chest and broke an ankle. Fortunately the wind was driving the bulk of the deadly cloud in the opposite direction.

*

Blissfully unaware of the tragedy occurring a few hundred yards from the Kali Grounds, Dilip and Padmini's wedding guests innocently gave themselves up to celebration. Padmini had kept a surprise in store for them. No feast took place in India without homage also being paid to the gods. That night the young woman was going to give thanks to Jagannath for all his blessings by dancing for him and for all the occupants of Orya Bustee. Discreetly she had gone to her hut to change from her wedding clothes into the costume worn by performers of *Odissi*, the traditional Orissan dance. True, it was not made out of silk embroidered with gold thread like those of the temple dancers, but of simple cotton material; but what did that matter? Dalima and Sheela adjusted her bodice and draped the material round her thighs and legs before spreading it out like a fan from her waist to her knees. Then they caught the young woman's long tresses up in a bun and adorned it with a jasmine flower. Then they fastened imitation jewellery round her neck, in her ears, round her arms, wrists and waist. Finally they put anklets with bells on round her ankles. The god would be pleased. The blood of Orissa definitely flowed through the veins of the former peasant girl from Mudilapa. And it was the thousand-year-old culture of her distant province that carried the young newly-wed along, as her bare feet began to pound the *mandap* on which she had but a short time previously sealed her marriage.

Dalima's singing and Dilip's staccato beating of two tambourines accompanied her dance. The crowd of enthralled guests cried out in delight: the men with *vah! vah*s! and the women with *you! you*s! that fired their poverty-stricken neighbourhood with triumphal fervour. Suddenly, however, Belram Mukkadam raised his stick above the audience. He had just heard the distant howl of Carbide's siren. Padmini's feet stood still, the bells on her ankles fell silent. Everyone strained their ears anxiously in the direction of the metal structure, which still appeared so peaceful in the distant halo of its thousand light bulbs.

'Don't say we're going to have to go through what we did the other evening,' Prema Bai protested vehemently. 'Because I, for one, am staying at home this time.'

Yet again it was Rahul who allayed their fears.

'You're getting uptight about nothing, friends,' he assured them. 'Since the last alert, they've decided to demolish their factory. But apparently it's so rotten they're frightened they won't be able to dismantle it. It's riddled with holes.'

'Perhaps that's why the siren's going, like the other evening when there was a gas leak,' suggested the dairyman Bablubhai.

His remark went unanswered: the howl of the siren had suddenly stopped. Padmini started to dance again, Dalima resumed her singing and Dilip his tambourine beating. The show went on even more enchantingly than before. The god was being really indulged. And the guests too. But why could they no longer hear the siren? None of them knew that those in charge of the factory had recently modified it. In order to make it easier to broadcast instructions to the workers during an emergency and to prevent the neighbours from panicking at the least little incident, the siren automatically stopped after ten minutes. A quieter alarm, which could not be heard outside the factory boundaries, took over.

Soon, however, there were other indications to arouse the anxious curiosity of the revellers. First it was a pungent odour.

'Little mischief-makers throwing chillies on the *chula* again!' suggested Ganga Ram who, as a former leper, had a particularly keen sense of smell.

'Bah!' replied Iqbal, 'you know very well that it's tradition . . .'

He was interrupted by an ear-splitting bellow. Out of the darkness surged Nandi the bull with his painted horns and the five cows Mukkadam and his friends had bought with Carbide's compensation money, staggering as if they were drunk. They were vomiting yellow froth, their pupils had swollen up like balloons and great burning tears poured from their eyes. The animals took a few more steps, then sank to the ground with a last rattle. It was one-thirty in the morning. On the Kali Grounds, the apocalypse had begun.

⁂

The two geysers of gas had merged to form an enormous cloud about a hundred yards wide. Twice as heavy as air, the MIC made up the base of the gaseous ball that was formed by the chemical reaction in tank 610. Above it, in several successive layers, were other gases, among them phosgene that had escaped from a nearby reactor, hydrocyanide acid and monomethylamine with its suffocating smell of ammonia. Because these gases were lighter in density the cloud would spread more rapidly, more widely and further. At the same time the movement of the noxious bank of fog was not homogenous. It progressed in fits and starts, striking or sparing according to the temperature of the location, the degree of humidity and the strength of the wind.

The vapours that reached the areas closest to the factory poisoned in a random fashion along the way, but the smell of boiled cabbage, freshly cut grass and ammonia covered the entire area in a matter of seconds. No sooner had Belram Mukkadam spotted the cloud, than he felt its effects. Realizing that death was about to strike, he yelled: '*Bachao! Bachao!* Get out of here!' The wedding guests were immediately seized with panic and ran off in all directions.

For Bablubhai, it was already too late. Orya Bustee's dairyman would never again bring milk to children suffering from rickets. When Nandi the bull died, he left the banquet and rushed to his stable where he could hear his buffalo cows bellowing. The seventeen beasts were lying down when they were hit head-on by a small blanket of gas moving along at ground level. Several had already succumbed. Devastated, Bablubhai ran to his hut to save his newborn son and his wife Boda.

'The oil lamp has gone out,' murmured the young woman tearfully.

Bablubhai bent over to grab his child. A gust of vapour caught him there. It paralysed the dairyman's breathing instantaneously and he was struck down in a faint, over the body of his lifeless baby.

Similar respiratory paralysis overtook several of the other guests in mid-flight. Another small greenish cloud laden with hydrocyanide acid drifted into old Prema Bai's hut. It killed the midwife outright, as she lay on her *charpoy*. She and many of the other guests had sought refuge in their homes. In the hut next door, Prodip and Shunda, Padmini's grandparents, also succumbed in seconds. Of all the gases making up the toxic mass, hydrocyanide acid was one of the deadliest. It blocked the action of the enzymes carrying oxygen from the blood to the brain, causing immediate brain death.

One of the first victims of this creeping layer of gas was the cripple Rahul on his plank on wheels. Because of his robust constitution, he did not die straight away but only after several minutes of agony. He coughed, choked and spewed up blackish clots. His muscles shook with spasms, his features contorted, he tore off his necklaces and his shirt, groaning and gasping for something to drink, then finally toppled from his board and dragged himself along the ground in a last effort to breathe. The man who had always been such a tireless source of moral support to the community, who had so frequently appeased the fears of his companions in misfortune, was dead.

Awakened with a start by all the yelling and shouting, those who were not at the wedding and were asleep, rushed panic-stricken out of their huts. For the first time Muslim women emerged with their faces uncovered. From out of all the alleyways came small carts laden with old people and children. Very soon, however, the men pulling them suffocated and collapsed. Unable to get back on their feet, they lay sprawled out in their own vomit. Little girls and boys who were lost, fastened on to passing fugitives and bicycles. Many of the residents of Chola and Jai Prakash Nagar Bustees took refuge in the small temple of Hanuman, the monkey god, or in the little mosque that was soon overflowing with distressed people. In their panic, men and women left other family members behind in their huts. Ironically their activity was often their undoing, while those left behind were in many instances recovered alive. The gas claimed more victims amongst those obliged to breathe deeply because they were moving, than those who kept still.

Others, such as Iqbal and Bassi, made sure, before they fled, that no one was left behind in any of the lodgings in their alleyway. That was how they came to find the old mullah with his goatee beard in one of them. Persuaded that Allah had decreed that the world should end that night, the holy man had knelt down on his prayer mat and was reading *suras* from the Koran by the light of a Carbide lamp.

'You are my creature, and you will not rise up against My will,' he repeated as his neighbours scooped him up to carry him away. As he emerged from the hovel, into which the deadly vapours were about to pour, he enquired of his rescuers: 'Are you quite sure that the end of the world is tonight?'

In the fetid, stinking darkness people shouted for their spouses, children or parents. For those blinded by the deadly emissions, shouting a name became the only way of making contact with their loved ones again. Time and again Padmini's name resounded through the night. In the stampede, the heroine of the evening had found herself brutally separated from her husband, mother and brother. She too was almost blind. Carried along by the human torrent, with her bells jangling round her ankles, coughing blood, Padmini did not hear the voices calling out to her. And soon the calling stopped: people's throats had constricted from the gas, lungs were choked and no one could utter a sound. In an effort to relieve the dreadful pains in the thorax, the poor wretches were squeezing their chests with all their strength. Stricken with pulmonary oedema, many of them coughed up a frothy liquid streaked with blood. Some of the worst affected spewed up reddish streams. With their eyes bulging out of their heads, their nasal membranes perforated, their ears whistling and their cyanosed faces dripping sweat, most of them collapsed after a few paces. Others, overcome with palpitations, dizziness and unconsciousness fell right there in the doorways of the huts they had tried to leave. Yet others suddenly turned violet and coughed dreadfully. Their coughing resounded through the night in sinister harmony.

Amidst all this chaos, one man and one woman walked, with difficulty, against the tide. Having given the signal for everyone to escape, Belram Mukkadam had decided to go in the opposite direction. He was taking his wife Tulsabai back to their hut. The mother of his three children wanted to die at home. Suffering from awful stomach pains, unable to breathe any more, the poor woman stumbled over the corpses that lay outstretched in the alleyways. On arriving outside her hut, she turned round to look for her husband. It was then that she realized that the last body she had tripped over was Belram's. Half-blinded, she had not seen him fall. The pioneer of Orya Bustee, the man who had drawn out the plot for each of its huts with the tip of his stick, who for twenty-five years had protected the poor, restored their dignity and fought for their rights, the legendary manager of the tea house, had, in his turn, been brought down by Carbide's gas.

Many of the bustee dwellers believed doors and windows could keep out the gases. They tried to take refuge in brick built houses. The nearest one was the godfather Omar Pasha's. With its two storeys of brickwork, it rose out of the disaster area like a fortress. Persuaded that the blanket of gas moved along the ground, the old man had retreated to the second floor with his family and his best fighting cocks. In the panic, Yagu, winner of that Sunday's duel, had been forgotten. Brought down by the toxic gases, he lay with burst lungs in the living room on the ground floor.

The godfather had his servants and bodyguards take in the refugees. Their arrival was greeted with acts of extraordinary generosity. Omar Pasha's eldest son took a little girl who was hardly breathing in his arms and laid her gently on the *charpoy* in his bedroom. The women of the house immediately removed their muslin veils and set about soothing people. They dipped the pieces of material in a bowl of water and applied them as cooling compresses to eyes that were on fire. One of Omar's wives, a plump matron whose arms jangled with bracelets, sponged away the blood flowing from people's lips, handed out glasses of water, and comforted one and all. Even Omar Pasha himself helped. With gold-ringed fingers he handed round plates of biscuits and sweets, a kindness that the survivors of that apocalyptic night would never forget.

Not all the brick built houses bordering on the slums were as welcoming. Ganga Ram and Dalima chose to flee along the railway line leading to Bhopal station. Ganga was convinced that he would find refuge a little further on in one of the villas occupied by railway executives. He knocked on the door of the first but received no response. Afraid that the wave of gas would catch up with him again, he did not hesitate to break a window pane to climb inside. At this point there was a series of gunshots. Believing he was the victim of a burglary and still unaware of the accident at the factory, the owner of the property had fired his revolver. Fortunately, in the darkness, he missed his target.

⁕

The horrific, the unspeakable, was happening. Driven by the wind, the wave of gas was catching up with the flood of humanity trying to escape. Out of their minds with terror, people were running about in all directions, with shredded clothes and torn veils, trying to find a pocket of breathable air. Some, whose lungs were bursting, rolled on the ground in awful convulsions. Everywhere the dead with their greenish coloured skins lay side by side with the dying, still wracked with spasms and with yellowish fluid coming out of their mouths.

Amidst this hell, the bicycle repairman Salar came upon

a vision that would haunt him. As he reached the corner of Chola Road, he narrowly escaped being knocked over by a white horse, bridled and saddled as if for some celebration. Through the veil of gas burning his eyes, he recognized the white mare which only a few hours earlier Dilip, Padmini's bridegroom, had ridden to his wedding ceremony. With its eyes bloodshot, its nostrils steaming with burning vapours and its mouth foaming with greenish vomit, the animal bolted away, came back at a gallop, stopped sharp, gave a heart-rending whinny and collapsed.

Of all the extraordinary scenes that marked that night of horror, one in particular would leave an impression on the few survivors: the frantic flight of a fat man in his underpants and vest, gasping his lungs out behind a heavily loaded cart. Nothing could have prevented the usurer Pulpul Singh from taking with him something more precious than life itself: his safe full of bank notes, jewels, watches, transistors, gold teeth, and, above all, the property deeds pawned by the residents of Orya Bustee.

# 40

## 'Something beyond all comprehension'

Less than 400 yards from the apocalypse taking place on the Kali Grounds, a stout man fiddled happily with his moustache. Sharda Diwedi had won. None of his power station turbines had failed. Bathed in an ocean of light, the marriage ceremony of his niece Rinou was passing off with all the brightness hoped for. The final part of the ritual was reaching its conclusion. The girl's father was waiting for a signal from the officiating priest to pronounce the words which would officially seal the union of bride and groom. 'I give you my daughter, in order that my one hundred and one families may be exalted for as long as the Sun and the Moon continue to shine, and with a view to having an heir.' The guests assembled under the beautiful *shamiana* hired from Parvez held their breath. In a few seconds' time, these words would bind the two young people together for ever. But they were never to be uttered. The ceremony was brutally interrupted by shouting. 'There's been an accident at Carbide's! *Bachao!* Get out of here!' frantic voices yelled from all directions.

Already a suffocating smell was invading the heart of the

Railway Colony. Moving in small pockets at different heights, the cloud seeped round the buffet tables, the dance floor, the swimming pool, the musicians' stand, and the cooks' braziers which immediately flared up in a chemical reaction. As dozens of guests collapsed, the chief stationmaster Harish Dhurve was hit by deadly vapours. Letting go of his last glass of English liquor, he fell to the ground. Dr Sarkar, who had forbidden him any alcohol, braved the blanket of toxic gas and tried to resuscitate him, to no avail. A few minutes before Bhopal station was hit, it lost its stationmaster.

Panicked, Sharda Diwedi tried to telephone the only person he believed was in a position to explain what was going on. But Jagannathan Mukund's telephone line was continuously engaged. In between two attempts, Diwedi's own telephone rang. He recognized the voice of the man in charge of the electricity substation in Chola.

'Sir, we're surrounded by a suffocating cloud of gas. We're requesting permission to leave. Otherwise we're all going to die.'

Diwedi thought briefly.

'Whatever you do, stay right where you are!' he urged. 'Put on the masks Carbide gave you and block up all doors and windows.'

'Sir,' replied the voice, 'there's just one problem: there are four of us and only one mask.'

Disconcerted, Diwedi searched for the right thing to say.

'You'll just have to take it in turns,' he eventually advised.

At the other end of the line there was a derisive laugh and then a click. His employee had hung up. The head of Bhopal's power station had no idea that he had just saved four men's lives. Next day when the military collected up the dozens of corpses sprawled about the approaches to the substation, they would be surprised to discover four workers inside, still breathing.

'*Bachao! Bachao!*' Coughing, spitting, suffocating and with burning eyes, Rinou and her fiancé found themselves caught up in a nightmare, along with all those who had come to celebrate with them. They were scrambling about in all directions, desperate for something to drink, fleeing in the direction of the station, seeking refuge in the local houses. Realizing that the panic-stricken crowd must be evacuated before the cloud killed everybody, Diwedi overcame the bout of coughing that was setting his throat on fire and ran to the garages to requisition the trucks belonging to the *shamiana* hirer and the caterer. But the garages were empty. Even his own car had disappeared. At the first cries of *Bachao!*, the cooks, servants, the men who put up the tents, the electricians and musicians had all jumped in the vehicles and driven off. The four men in charge of the generator set had decamped on their scooters. The indomitable little man decided then to go on foot to his home, 700 or 800 yards away, where he would at least find his old Willis jeep. On his way back he was intercepted by a frenzied crowd. People stormed his old banger, throwing themselves onto the seats, bonnet and bumpers. There were twenty, thirty, fifty of them, struggling with the last vestiges of their strength to climb on board. These were the survivors from the Kali Grounds' neighbourhoods. They were weeping, pleading, threatening. Many of them, exhausted by this final effort, collapsed unconscious. Others brought up the last blood from their lungs and keeled over. A truck then roared like a rocket through the crowd of dying people. Diwedi heard skulls cracking against the bumpers, radiator and wings. The driver left a pulp of crushed bodies in his wake before disappearing. A moment later, Diwedi made out through vapour-burnt eyes, a woman throwing her baby over the guard rail of the bridge over the railway line, before jumping into the void herself. 'I realized then that something awful was going on,' he would later say, 'something beyond all comprehension.'

※

The Reverend Timothy Wankhede had spent his Sunday afternoon preaching to hospital patients on the epistle of St Paul, imploring the mercy of the Lord upon his children, who in the pursuit of riches had 'fallen into temptation and a snare, and into many hurtful and foolish lusts, which drown men in destruction and perdition.' The young rector and his wife Sobha had just been woken with a start by the cries of Anuradh, their 10-month-old son. The toxic vapours had entered the modest red-brick vicarage they occupied in the Railway Colony, next door to The Holy Redeemer Church. In a few seconds they too were overtaken by the same symptoms that were affecting all the other victims of that hellish night. They struggled to understand what was going on.

'Perhaps it's an atomic bomb,' Timothy ventured with difficulty, his throat was burning.

'But why in Bhopal?' asked Sobha, discovering, to her horror, that blood was trickling from her baby's lips.

Her husband shrugged his shoulders. He knew that he was going to die and was resigned to it. But as a man of God and despite his pain, he wanted to prepare himself and his family for death.

'Let's pray before we leave this world,' he said calmly to his wife.

'I'm ready,' replied the young woman.

Timothy made an effort to stand up, took his child in his arms and led his wife over to the other side of the courtyard. He wanted to spend his last moments in his church.

Placing the small body on a cushion at the foot of the altar, he went and fetched the bulky copy of the New Testament from which he liked to read each week to his parishioners, and came back to kneel beside his wife and child. He opened the book at Chapter 24 of the Gospel according to St Matthew and recited as loudly as his burning throat would permit. 'Watch therefore: for ye know not what hour your Lord doth come . . .' Then they drew consolation from the words of the psalmist. 'Yea, though I walk through the valley of the shadow of death, I will fear no evil,' Timothy read with feeling.

Suddenly, through the stained glass of the small church, there appeared the figure of a saviour. With a damp towel plastered over his nose and mouth, Dr Sarkar signalled to the priest and his wife to protect themselves in the same way, and come out immediately. There were already five people piled into the doctor's Ambassador waiting outside the church, but in India there was nothing unusual about that. Timothy Wankhede, who had found Jesus Christ while listening to the radio one day, took time to put a bookmark at the page containing Chapter 24 of St Matthew's Gospel. Despite the calvary he had endured, which would leave both him and his family with serious after-effects, his 'hour' had not arrived.

❉

'Your samosas are great!' enthused Satish Lal, the luggage porter.

He and his friend Ratna Nadar were waiting at the end of platform no 1 for the Gorakhpur Express. Like the ninety-nine other coolies, Lal had polished off the contents of the small cardboard box brought by Padmini's father.

'They certainly all seemed to have tucked in,' said Nadar, proud to have been able to give his friends a treat.

'I'll bet you're going to have to tighten your belt a bit now,' observed Lal. 'I can't imagine Pulpul Singh giving anything away.'

'You can say that again!' confirmed Nadar.

All of a sudden the two men felt a violent irritation in their throats and eyes. A strange smell had just invaded the station. The hundreds of passengers waiting for their trains also felt their throats and eyes become inflamed.

'It's probably an acid leak from one of the goods wagons,' said Lal, who knew that there were containers of toxic material waiting to be unloaded. 'It wouldn't be the first time!'

Lal was wrong. The toxic cloud from the factory had arrived. It would turn the station into a deathtrap for thousands of travellers.

※

The two coolies rushed to the stationmaster's office at the end of the platform. The deputy stationmaster V. K. Sherma was just moving one of the pins on the traffic indicator board. The Gorakhpur Express was approaching Bhopal. It was due to arrive in twenty minutes.

Lal could scarcely speak.

'Boss,' he croaked, 'something's going on . . . people on the platform are coughing their guts out. Come and see!'

The deputy stationmaster and his assistant Paridar left the office but were immediately hit in the face by a pocket of poisonous gas moving at head height. Two or three inhalations were enough to stop any air reaching their lungs. With their ears whistling and their throats and faces on fire, they beat a retreat, gasping for breath.

Witnessing the scene, the young traffic regulator Rahman Patel had the presence of mind to do the only useful thing possible. He closed all apertures and turned on the air conditioning. The gusts of fresh air it emitted brought immediate relief to the two railway employees who slowly regained their senses. That was when the internal line telephone rang. Sherma recognized the voice of the man in charge of the Nichadpura Centre, a fuel depot a few hundred yards from the Carbide factory.

'There's been an explosion at Carbide,' announced the panic-stricken speaker. 'The whole area is covered with a toxic cloud. People are scrambling about in all directions. Get ready. You'll be hit next. The wind is blowing the cloud in your direction . . .'

'It's already here,' replied Sherma.

A vision of horror passed through the deputy stationmaster's mind at that moment: the Gorakhpur Express was speeding towards Bhopal with hundreds of passengers on board.

'Whatever happens we've got to make sure the train doesn't stop,' he cried to his two assistants.

No sooner had he spoken, however, than he shook his head. He knew what Indian railways' bureaucracy was like. An initiative like that could not be taken at his level. Only the chief stationmaster could issue such an order. Sherma immediately dialled Harish Dhurve's home. No one answered.

'He must be downing a last whisky at the Railway Colony wedding,' he remarked, frustrated.

There was little point in trying again. He could never receive the necessary authorization. His boss had been dead for half an hour.

❋

There were no vendors, lepers, beggars, coolies, children or travellers left. Platform no 1 was nothing more than a charnel-house of entangled bodies, stinking unbearably of vomit, urine and defecation. Weighed down by the gas, the toxic blanket had draped itself like a shroud over the people chained to their baggage. Here and there, the odd survivor tried to get up. But the deadly vapours very quickly entered their lungs and they fell back with mouths contorted like fish out of water. The beggars and lepers whose tubercular lungs were already weak, had been the first to die.

Thanks to the air conditioning filtering their air, the three men in the stationmaster's office and a few coolies who had taken refuge in their cloakroom had so far managed to escape the noxious fumes. In vain V. K. Sherma frantically cranked his telephones to call for help. All the lines were engaged. At last he managed to speak to Dr Sarkar. After evacuating the priest and his family, the railway workers' doctor had gone back to his chambers in the Railway Colony. From behind the damp compresses over his mouth and nose, he sounded confused. He had just spoken to Dr Nagu, Director of the Madhya Pradesh Health Service.

'The director was furious,' said Sarkar. 'He told me the people at Carbide didn't want to reveal the composition of the toxic cloud. He tried to insist and asked whether they were dealing with chlorine, phosgene, aniline or I don't know what else. It was no use. He wasn't able to find out anything. He was told the gases were not toxic and that all anyone had to do to protect themselves was put a damp handkerchief over their nose and mouth. I've tried it and it seems to work. Oh! I was forgetting . . . they also told the director to: "breathe as little as possible!" My poor Sherma, pass that advice on to your travellers while they're waiting for help to arrive.'

Help! In his station strewn with the dead and the dying, the deputy stationmaster felt like the commander of a ship about to be engulfed by the ocean. Even if he could do nothing for the passengers on platform no 1, however, he must still try and save those due to arrive. Unable to contact his boss to prevent the Gorakhpur Express from stopping at Bhopal station, he would still do all he could to prevent it from running into the trap. The only way was to halt it at the previous stop. His assistant immediately called the station at Vidisha, a small town less than twelve miles away.

'The train has just left,' the stationmaster there informed him.

'Curses on God,' groaned Sherma in consternation.

'Is there at least a signal we could switch to red?' asked Patel, the young traffic regulator.

The three men looked at the luminous indicators on the large board on the wall.

'There isn't a single point or signal box between Vidisha and Bhopal,' Sherma established.

'In that case, we'll just have to run out in front of the train and signal the engine driver to stop,' declared Patel.

The idea appeared to stupefy his two elder colleagues.

'And how are you going to signal to an engine driver to stop a train going at full speed in the middle of the night?' asked Sherma's assistant.

'By waving a lamp about in the middle of the track!'

Sherma nearly swallowed his *pan*. The whole idea seemed outrageously dangerous. But after a few seconds he changed his mind.

'Yes, you're right, we could stop the train with lanterns. Go and fetch some able-bodied coolies.'

'I'm volunteering,' announced Patel.

'So am I,' said Paridar, Sherma's assistant.

'Okay, but it will take at least four or five of you. Four or five lanterns will be easier to see in the dark.'

Patel rushed to the washbasin at the far end of the room to soak his *gamcha*. After wringing it out, he plastered it over his face and went out. Two minutes later he came back with Padmini's father and his friend Satish Lal, who had both escaped the gas attack by taking refuge in a first class waiting room and shutting the windows. Sherma explained their mission to them, emphasizing how vital it was.

'If you can stop the Gorakhpur, you may save hundreds of lives,' he told them. Then he added: 'You'll be heroes and be decorated for it.'

The prospect brought only the faintest of smiles to the four men's faces. Sherma pressed his hands together over his chest.

'May God protect you,' he said, inclining his head. 'You'll find some lanterns in the maintenance store. Good luck!'

The deputy stationmaster was overcome with emotion. 'Those men,' he thought, 'are real heroes.'

Guided by Padmini's father, who knew every turn of the track by heart, the little procession moved off into a murky darkness filled with invisible dangers. Every five minutes Ratna Nadar would raise one arm to stop his comrades, kneel down between two sleepers and press his ear to one of the rails. There was as yet no vibration from the approaching train.

✳

Huddled with her two sons on the seat of one of the forty-four carriages, Sajda Bano was counting off the last minutes of her interminable journey back to the city where her husband had been Carbide's first victim. When she felt the train slow down, she moved nearer to the window to gaze out at the illuminated outline of the factory that had put an end to her happiness. She had dreaded the return to Bhopal, imposed by her in-laws' determination to get their hands on the 50,000 rupees' compensation awarded by the factory.

Sajda had experienced all the hardship of being an Indian widow. No sooner had her husband been buried than her father-in-law had thrown her out of the house, on the pretext that she was refusing to renounce her inheritance. Out of her mind with grief and despair, the young widow had responded with her first act as an independent woman. She had torn off the veil she had worn since she was nine years old and rushed to the bazaar to sell it. The 120 rupees she received in return were the first money she had ever earned. Since then she had never again worn a veil. Overcoming the triple handicap of being a woman, a Muslim and a widow in a country where, despite all the progress, customs could still sometimes be medieval, she embarked on a struggle for justice. She knew that she could count on the support of the kindly H. S. Khan, a colleague of her husband's, who had taken her and her children in after her in-laws had put her out on the street. She had stayed with him while she looked for lodgings and engaged a lawyer. Now she very much hoped that Khan would be on the platform to greet her. Poor Sajda! Having killed her husband, Carbide's gas had just struck down her benefactor on his way to the station.

❋

Holding their lanterns at arm's length, the four men progressed with difficulty. Without realizing it, they were passing through a multitude of small residual clouds, still lurking between the rails and along the ballast. They stumbled over corpses knotted into horrible attitudes of pain. Here and there, they could hear death rattles, but there was no time to stop. Then a great roar rent the darkness, accompanied by the same shrill whistle that made the occupants of the Kali Grounds tremble in their sleep. The train! . . .

Brandishing their lanterns, the four men rushed to meet it. Very swiftly, however, they ran out of breath. In the end the toxic vapours had penetrated their damp cotton compresses. Hyperventilating with the effort, their lungs craved more and more air, the same air that was poisoned with deadly molecules. The weight of their lanterns became unbearable. And yet they kept on going. Staggering between the sleepers, suffocating and vomiting, the four men desperately waved their lights, but the engine driver of the Gorakhpur Express did not understand the signal. Thinking they were revellers fooling around by the railway track, he kept on going. By the time, in a horrifying flash, he saw the men yelling at him from the middle of the rails, it was too late. With its engine cowling spattered with flesh and blood, the train was already entering the station.

❄

The headlights of the locomotive surging out of the mist made the deputy stationmaster jump. V. K. Sherma realized that his men's attempt had failed. The train glided smoothly along the rails of platform no 1 before stopping with a deafening grinding noise. There was still one last chance to prevent the worst.

Like all large stations in India, Bhopal was equipped with a public address system. V. K. Sherma dashed to his console at the far end of the office, turned on the system and grabbed the microphone. 'Attention! Attention!' he announced in Hindi with as calm and professional a voice as he could. 'Because of a leak of dangerous chemical substances, we invite all passengers due to get out at Bhopal to remain in their carriages. The train will depart again immediately. Passengers may get out at the next station, from where buses will transport them to Bhopal.' He repeated his message in

309

Urdu. He was very quickly able to gauge the success of his announcement. Doors were opening, people were getting out. Nothing could threaten the lives of pilgrims coming to celebrate Ishtema. They were secure in the knowledge that God would protect them.

With a wet towel over his mouth, Sherma left his post to run to the head of the train and order the engine driver to leave. He knew that this order was illegal. All trains stopping in Bhopal were required to undergo routine mechanical checks. Curtailing a stop meant preventing these checks. That night, however, there were no maintenance teams or parts supervisors left. There were only hundreds of people who might yet be saved. Terrified that the vapours might have reached the engine driver, that he might have passed out or already be dead at the controls of his locomotive, Sherma hurried as fast as he could. Recognizing his uniform, dying people clung to him in a last desperate effort. Others threatened him and tried to block his way, demanding help. Stepping over bodies and slipping in vomit, at last he reached the front of the train. There his railway worker's reflexes came back to him. He took his little flag out of his pocket and banged on the window of the locomotive's cab.

'All clear. Depart immediately!' he announced.

That was the ritual formula. The engine driver responded with a nod of his head, took the brakes off and leaned hard on the regulator of his diesel engine. Slowly the Gorakhpur Express extricated itself from the dreadful necropolis to the accompaniment of grinding noises and whistle blasts. Drenched in sweat, breathing painfully and with a pounding heart, but proud of his achievement, the deputy stationmaster picked his way back through the carnage to his office at the other end of the platform. But Bhopal's stationmaster's office was no longer recognizable.

The small notice 'A/C office' displayed over the door had

attracted some of the passengers driven frantic by the toxic cloud. In the conviction that the gases would be unable to get into an air-conditioned room, they had rushed in, ransacking everything in their path, breaking up the train indicator board, tearing out the telephones. Disaster reigned. Even the appearance of the tall figure of Dr Sarkar failed to calm the plunderers' fury. The railway doctor had managed to get to the station on foot. He was carrying a bag with a red cross on it, a derisory symbol in this setting of agony and death. He had filled his bag with bottles of eye lotion, cough lozenges, bronchodilators, throat pastilles, cardiac stimulants and anything else he could find in his medicine cabinet. But what use were such remedies? The doctor bent over the first body. Then, on the platform, he came across a scene that would haunt him for the rest of his life: a baby suckling at the breast of its dead mother.

❁

Like many other passengers on the Gorakhpur Express, Sajda Bano had not heard the deputy stationmaster's announcement. She got out with her two children and her suitcases. In the yellowish mist enveloping the platform, she tried to look for the figure of the good Mr Khan, her husband's friend. But with her eyes smarting from the vapours, she could only make out a confusion of corpses, in a deathly silence. 'It was as if the train had stopped in a cemetery,' she said. Three-year-old Soeb and five-year-old Arshad were immediately assailed by the gases and racked with coughing. Sajda herself felt her throat and trachea become inflamed. She could not breathe. Stepping over the corpses, she dragged her sons towards the waiting room in the middle of the platform. The room was filled to overflowing with people on the verge of death, coughing, vomiting, urinating, defecating and delirious. Sajda stretched the two boys out in a corner of a seat, put a teddy bear, a gift from their grandmother, in the youngest's arms, and placed two wet handkerchiefs over their little livid faces. 'Don't worry,' she told them, 'I'm going to get help and I'll be back straight away.' As she went out, she passed the window to the ticket sales and reservations office. The portly Mr Gautam looked as if he was sleeping. His lifeless head was propped on a pile of registers.

All night long Sajda Bano wandered about amongst thousands of Bhopalis, looking for a vehicle to come and take her children to hospital. The panic in the station and surrounding area was such that she did not get back to them until the early hours of the morning. She found her two boys where she had left them. Little Soeb was still clutching his teddy bear to his chest and breathing weakly, but clotted blood had formed a red ring around the motionless lips of his brother Arshad. Sajda knelt down and put her ear to the frail, lifeless chest. Carbide's gas had taken her husband. Now it had stolen one of her children too.

# 41

## 'All hell has broken loose here!'

It was a silent, insidious, and almost discreet massacre. No explosion had shaken the city, no fire had set its sky ablaze. Most Bhopalis were sleeping peacefully. Those still revelling in the reception rooms of the Arera Club, under the wedding *shamianas* of the rich villas in New Bhopal, or in the smoke-hung rooms of Shyam Babu's restaurant, overrun that night, as every Sunday night, with medical college students – all those people suspected nothing. In Spices Square in the old city, an exultant crowd went on acclaiming the *mushaira*'s poets. Salvos of ecstatic *vah! vah!*s shook the window panes of the neighbourhood. Even the eunuchs had turned out in force, a rare thing, because it was one of their rules to be home by sunset. The presence of the legendary Jigar Akbar Khan, however, and of several other masters of poetry from all four corners of the country, had persuaded the gurus of the various eunuch 'families' to give their protégés free reign. There was just one condition: they must travel in groups of four. The audience contained some of the more famous members of their unusual community: the plump Nagma, for example, the ravishing Baby and the disconcerting Shakuntual with her large, dark, kohl-encircled eyes.

In accordance with tradition, the *mushaira* also gave a few unknown amateurs the opportunity to recite their poetry. The Muslim workman who, until 23.00 hours, had been busy flushing out the pipes in the Carbide factory, was among those privileged few. When his turn arrived, however, Rehman Khan froze with fright. His young son Salem took his hand and led him onto the stage. The crowd held its breath. The hands that had just set off an inevitable tragic sequence gripped the microphone.

> *Oh my friend, I cannot tell you*
> *Whether she was near or far,*
> *Real or a dream . . .*

said the worker-poet fervently, his eyes half-closed.

> *It was like a river flowing through my heart.*
> *Like a moon lit up, I devoured her face*
> *And felt the stars dance about my head . . .*

✳

Jagannathan Mukund would not go picnicking with his son beside the Narmada's sacred waters the next day. The sound of his telephone ringing had just rudely awoken the Managing Director of the factory where Rehman Khan worked. S. P. Chowdhary, his production manager, informed him that a gas leak had occurred in the MIC storage zone. Mukund refused to believe it. He simply could not let go of the idea that an accident could not happen in a factory that was stopped.

'Come and get me,' he ordered Chowdhary. 'I want to go and look at the site.'

While he was getting dressed, the telephone rang again. It was Swaraj Puri, the city's police chief, to inform him that panic-stricken residents were fleeing from the Kali Grounds. Many of them showed signs of poisoning. Mukund decided to call his friend, Professor N. P. Mishra, Dean of the Gandhi Medical College and Chief of Internal Medicine at Hamidia Hospital. The doctor had just come back from a wedding.

'N. P.!' he warned, 'get ready for some emergency admissions at the hospital. It seems there's been an accident at the plant.'

'Is it serious?' asked Mishra anxiously.

'I'm sure not, the factory's out of production. A few inconsequential poisonings, I imagine.'

'A gas leak?'

'So they tell me. I'll know more when I've visited the scene.'

The doctor pressed his friend.

'Phosgene?' he asked, remembering the death of Mohammed Ashraf.

'No, methyl isocyanate.'

This answer left the professor at a loss. Carbide had never supplied Bhopal's medical teams with any detailed information about the substance.

'What are the symptoms?'

'Oh, nausea, sometimes vomiting and difficulty in breathing. But with damp compresses and a little oxygen everything should be all right. Nothing really serious . . .'

Was this reputable engineer, chosen by Carbide to succeed the 'beautiful plant's' last American pilot, acting a part? Or was he simply ignorant? Did he really not know that MIC was a deadly substance? When, a few minutes later, he reached Hamidia Road, his white Ambassador was suddenly swamped by a throng of poor wretches coughing their lungs out, vomiting, groping their way about. Fists banged on the body of his car.

'Where are you going?' shouted a man who was frothing at the mouth.

'To the factory!' answered Mukund through the closed window.

'To the factory! You're mad! Turn back or you're dead!'

At these words, the engineer wound down his window. A powerful smell of chemicals overwhelmed the interior. The driver immediately started to choke. Crumpled over his steering wheel, he began to turn the car round.

'We've had it, Sir,' he wailed.

Mukund grabbed him by the arm.

'Carry straight on,' he ordered, pointing to the avenue leading up to Carbide's site. 'That's where we're going.'

Fortunately, Mukund had taken the precaution of bringing some handkerchiefs and a bottle of water. He handed out compresses to the production manager and the driver while the car carved its way through the middle of the fleeing crowd.

❊

In a matter of minutes the emergency department of Hamidia Hospital looked like a morgue. The two doctors on duty, Deepak Gandhe and Mohammed Sheikh, had thought they were going to have a quiet night after Sister Felicity's visit. All at once the department was invaded. People went down like flies. Their bodies lay strewn about the wards, corridors, offices, verandas and the approaches to the building. The admissions nurse closed her register. How could she begin to record the names of so many people? The spasms

316

and convulsions that racked most of the victims, the way they gasped for breath like fish out of water, reminded Dr Gandhe of Mohammed Ashraf's death two years earlier. The little information he could glean confirmed that the refugees came from areas close to the Carbide factory. So all of them had been poisoned by some toxic agent. But which one? While Sheikh and a nurse tried to revive the weakest with oxygen masks, Gandhe picked up the telephone. He wanted to speak to his colleague Loya, Carbide's official doctor in Bhopal. He was the only one who would be able to suggest an effective antidote to the gas these dying people had inhaled. It was nearly two in the morning when he finally got hold of Loya. 'That was the first time I heard the cruel name of methyl isocyanate,' Dr Gandhe was to say. But just as Mukund had been earlier, Dr Loya turned out to be most reassuring.

'It's not a deadly gas,' he claimed, 'just irritating, a sort of tear gas.'

'You are joking! My hospital's overrun with people dying like flies.'

Gandhe was running out of patience.

'Breathing in a strong dose may eventually cause pulmonary oedema,' Dr Loya finally conceded.

'What antidote should we administer?' pressed Gandhe.

'There is no known antidote for this gas,' replied the factory's spokesperson, without any apparent embarrassment. 'In any case there's no need for an antidote,' he added. 'Get your patients to drink a lot and rinse their eyes with compresses steeped in water. Methyl isocyanate has the advantage of being soluble in water.'

Gandhe made an effort to stay calm.

'Water? Is that all you suggest I do to save people coughing their lungs out!' he protested before hanging up.

He and Sheikh decided nonetheless to follow Dr Loya's

advice. Water, they found, did ease the irritation to the eyes and the coughing fits temporarily.

The situation into which the two doctors were brutally plunged was more horrific than any war story or tragedy they might have read about. 'What I liked more than anything else about my profession was being able to relieve suffering,' Gandhe would later say, 'and there I was unable to do that. It was unbearable.'

Unbearable was the fetid, foul breath from mouths oozing blood-streaked froth. Unbearable was the stupor in people's expressions, their inflamed eyes about to burst, their drawn features, their quivering nostrils, the cyanosis in their lips, ears and cheeks. Many of their faces were livid. Their discoloured lips already heralded death. Through their stethoscopes the two doctors picked up only the faintest, irregular sounds of hearts and lungs, or sputtering, grating, gurgling rattles. What struck them most was the state of torpor, bewilderment, exhaustion and amnesia in which they found most of the victims, which suggested that the nervous system had been profoundly affected.

Scenes of terror. A man and a woman broke through the crowd and laid their two children, aged two and four, on the examination table. Their heartbeats were scarcely perceptible and both were frothing at the mouth. Gandhe at once injected them with Derryfilin, a powerful bronchiodilator, bathed their eyes with lotion and gave each an oxygen mask. The children stirred. Their parents were overjoyed, convinced they were reviving. Then the little bodies went rigid. Gandhe listened with his stethoscope and shook his head. 'Heart failure,' he mumbled angrily.

This was only the beginning of his night of horror. Quite apart from haemorrhaging of the lungs and cataclysmic suffocation, he found himself confronted with symptoms that were unfamiliar to him: cyanosis of the fingers and toes, spasms in the oesophagus and intestines, attacks of blindness, muscular convulsions, fevers and sweating so intense that victims wanted to tear off their clothes. Worst of all was the incalculable number of living dead making for the hospital as if it were a lifeboat in a shipwreck. This onslaught gave rise to particularly distressing scenes. Going out briefly into the street to assess the situation, Gandhe saw screaming youngsters clinging to their mothers' *burkahs*, men who had apparently gone mad, tearing about in all directions, rolling on the ground, dragging themselves along on their hands and knees in the hope of getting to the hospital. He saw women abandon some of their children, those they could no longer carry, in order to save just one – a choice that would haunt them for ever. In desperation, the young doctor decided to appeal to his old master, the man whom, a few minutes earlier, Mukund had woken to inform that an accident was likely to give rise to 'some emergency admissions' at the hospital.

'Professor Mishra,' he begged after painting an apocalyptic picture of the situation, 'come quickly! All hell has broken loose here!'

❄

His appeal mobilized a chain of events marked by remarkable efficiency and extraordinary self-sacrifice. Two of the principal people involved would remain unknown. Santosh Vinobad and Jamil Ishaq were the operators on duty at the city's central telephone exchange located on the second floor of the central post office, just opposite the great mosque. Decaying and antiquated, it reflected India's backwardness when it came to telecommunications. Madhya Pradesh had only two circuits for international calls, and only a dozen lines to handle all domestic communications. Those Bhopalis

319

fortunate enough to have a telephone had to go through the operators to make any intercity calls. The bell jangled and Jamil Ishaq plugged in his connection. As soon as he heard the person on the other end of the line say 'Hello!', he exclaimed:

'Professor Mishra! I can hear you.'

The doctor, who had just set up his command post in his office, just opposite the Hamidia Hospital emergency block, was too disconcerted to speak.

'I recognized your voice, Professor. God be with you! I'll give all your calls priority.'

Mishra thought to himself that the whole city must know about the catastrophe. He expressed his gratitude.

'Whatever you do, don't thank me, Professor. This is the very least I owe you. You operated on my gall bladder a few weeks ago!'

Resisting the temptation to laugh, Mishra blessed his former patient's gall bladder and at once gave him a series of numbers in Europe and the United States. Since Carbide was refusing to supply any, he would ask the World Health Organization in Geneva and Medilas in Washington for any information they might have on treating MIC poisoning. But it was still Sunday in Europe and America. It would be another ten hours before offices opened and Mishra could obtain his information. In the meantime he decided to alert all pharmacists in the area and have them immediately bring all their stocks of bronchiodilators, antispasmodics, eye lotion, heart medication, and cough syrup and pastilles. After that he set about getting his colleagues, the deans of the medical schools in Indore and Gwalior, out of bed. He asked them to gather up all available medicines in their sectors and dispatch them by plane to Bhopal. Finally, he rang those in charge of the various firms in Bhopal that used oxygen bottles. 'Bring us all your stocks,' he told them. 'The lives of twenty, thirty, possibly even fifty thousand people are at stake.'

Once he had finished this telephone offensive, Mishra decided to rally all the medical students. Most were asleep in their hostel behind the medical college, after celebrating in Shyam Babu's restaurant. Mishra would wake them himself. He climbed the stairs, and went along the corridors, banging on their doors.

'On your feet, kids!' he cried. 'Don't waste time getting dressed! Come just as you are, but come quickly! Thousands of people are going to die if you don't get there in time.'

Mishra would never forget the sight of those boys and girls scrambling wordlessly out of bed and running almost in their sleep to the hospital. Some demonstrated their heroism immediately. One of them bent over a child suffocating by the gaseous vapours. Without any hesitation, he applied his mouth to the child's and breathed air from his own lungs for long minutes. This shock treatment miraculously revived the little boy. But when the medical student stood up, Deepak Gandhe saw him turn suddenly livid and stagger. In snatching the child from death, he had inhaled the toxic gas from his lungs. It was he who was to die.

It was not enough that the dreadful fog had burnt people's bronchia, eyes and throats. It had also impregnated their clothes, hair, beards and moustaches with toxic emissions so persistent that the medics themselves ended up experiencing symptoms of suffocation. A swift injection with Derryfilin mixed with ten cubic centimetres of Decadron was usually enough to prevent any complications. The courage generally displayed, however, did not mean that there were not moments of weakness; one panic-stricken young doctor tore the oxygen mask off a dying man, clamped it to his own face and greedily took a few gasps before fleeing. Yet, he came back at daybreak and for three days and three nights was one of the mainstays of the emergency wards.

Suddenly, in the midst of all the chaos, Sister Felicity appeared. She had left little Nadia momentarily to rush to the carnage of the wards and corridors. There were so many bodies all over the place that she could not move without bumping into an arm or a leg. She was struck by the impossibility of distinguishing between the living and the dead. People's faces were so swollen that their eyes had disappeared. She volunteered her services. Deepak Gandhe put her in charge of one of the rooms where an attempt was being made to regroup the victims into families. Felicity bent over an old man who lay unconscious beside the body of a woman in a mauve sweater. Gently she stroked his forehead. 'Wake up, Granddad! Tell me whether your wife was wearing a mauve sweater,' she insisted. The poor man did not answer and Sister Felicity turned to another woman stretched out between two young children. Were they hers? Or did they belong to a third woman a little further on, the one with cotton pads on her eyes?

In that terrible place of death, the living had lost the power of speech.

✳

Professor Mishra knew that the invasion was only just beginning. The toxic cloud would continue to wreak havoc. Thousands, possibly even tens of thousands of fresh victims would keep on coming. It was urgent that the campus between the medical college and Hamidia be turned into a gigantic field hospital. How were they going to achieve so mammoth a task in the middle of the night? Mishra had an idea. Once again, he picked up the telephone and woke Mahmoud Parvez, the man who rented out *shamianas*, who was fast asleep in his recently built house in New Bhopal,

322

safe from the toxic gases. Mishra told him about the tragedy that had struck the city, then promptly added:

'I need your help. You must go and get all the *shamianas*, all the carpets, covers, furniture and crockery you hired out for yesterday evening's weddings and bring them outside Hamidia Hospital as fast as possible.'

Parvez showed no trace of surprise.

'You can count on me, Professor! Tonight, those in need can have anything I own.'

The little man then woke his three sons, called all his employees to arms, sent his trucks out to every site where he had delivered the accessories and trappings for wedding celebrations. He had the two enormous *shamianas* set up in the courtyard of the Taj ul-Majid taken down. Never mind Ishtema! That night Bhopal was suffering and his duty as a good Muslim was to help relieve it. He directed one of his sons to empty his warehouses of any armchairs, settees, chairs and beds, not forgetting the famous percolator because 'a good Italian coffee can do a fellow a power of good'.

Marvellous Mahmoud Parvez! As his staff brought his goods to the afflicted, he kept one mission for himself. It was he, and he alone, who would dismantle the jewel of his collection, the magnificent, venerable *shamiana* embroidered with gold thread, which he had hired out to his friend, the director of Bhopal's electric power station, for his niece's wedding. The task came very close to killing him. Asphyxiated by a pocket of gas floating along the ground, Mahmoud collapsed, unable to breathe. By some miracle, a rescue team picked him up. He was among the first to receive emergency treatment under one of his own tents.

❋

Barely 500 yards from the improvised hospital into which the gas victims were pouring by the hundred, a man in a red pullover, his face protected by a damp towel and motor-cyclist's goggles, came out of a small house in the old part of town, in the company of his young wife and her 15-year-old sister. All three straddled the scooter that was waiting, propped against the door. The journalist Rajkumar Keswani had been woken a few moments earlier by a strange smell of ammonia. He had closed the window without ever for one moment imagining that the smell might be an indication of precisely the catastrophe he had warned the city against. He had called the police headquarters.

'What's going on?' he asked.

'An accident at Carbide,' a voice strangled with anxiety answered him. 'A gas tank explosion. We're all going to die.'

From his window Keswani then saw people fleeing in all directions, and understood. Settling his two passengers on the scooter, he gripped the handlebars and set off like the wind towards the distant neighbourhoods of New Bhopal, out of reach of the gases from the factory.

# 42

## *A half-naked holy man in the heart of a deadly cloud*

An act of barbarity had broken him; the Carbide catastrophe would make him a hero. One month after discovering six members of his family burnt alive in reprisal for the assassination of Indira Gandhi, the Sikh colonel Gurcharan Singh Khanuja, commanding officer of the engineering corps in Bhopal, found himself confronted with yet another tragedy. That night, with nothing to protect him but fireman's goggles and a wet towel over his face, the officer had sprung to the head of a column of trucks to rescue 400 workers at a cardboard factory and their families, all of whom were surprised by the gas as they slept.

Having completed that rescue operation, the Colonel and his men returned to the danger zone, this time to search the Kali Grounds' neighbourhoods for any survivors. The corpse of the white horse from Padmini's wedding was blocking the entrance to Chola Road. With its hooves in the air, its body swollen with gas and its eyes bloodshot, the animal was still in its harness. The soldiers tied a rope round its fore-limbs and pulled it to one side. A little further on, the officer came across other vestiges of the festivities: on the small *mandap* stage, the brazier for the sacrificial fire, flames still flickering, gilded armchairs, the musicians' drums and dented trumpets, saucepans full of curry and rice, and even the generator set hired to light up what should have been the greatest moment in Dilip and Padmini's lives. Abandoned outside a hut, Khanuja now found the wedding presents, some cooking utensils, clothing and pieces of material. He picked up the parasol the groom had carried as he processed on his white mare. With military discipline he took the time to jot down an inventory of all the debris in a notebook. Then, stepping over the corpses littering the alleyways, he systematically inspected every dwelling. He had given his men the order to proceed in total silence. 'We were on the alert for the slightest sign of life,' he said. Now and then they would hear a moan, a groan, a cough, a child crying. 'Bodies had to be moved to ascertain which ones were still alive,' the officer recalled, 'but often we were too late. The crying had stopped. There was nothing left but the dreadful, frightening silence of death.'

In one hut, Khanuja found an elderly couple sitting calmly on the edge of a *charpoy*. They smiled at the officer as if they had been expecting him to visit. In the shack next door an entire family had been wiped out: the parents and their six children sprawled out on the beaten earth floor, their eyes bulging, and foam and blood frothing out of their mouths. The youngest had died sucking their thumbs. Khanuja had the elderly couple taken away by truck and went off in search of other survivors. On Berasia Road where men came to beg Carbide's *tharagars* for jobs, the ground was scattered with bodies, struck down in mid-flight. Suddenly the Colonel's attention was drawn to that of a very young woman whose ankles sparkled in the moonlight. He turned on his torch and established that she was wearing anklets with bells on them, with her hands and feet decorated in henna, her close-fitting bodice and cotton loincloth draped in a fan shape over her hips and thighs. The officer thought she looked like one of the sacred dancers he'd seen on television. A white jasmine flower had been tucked in her bun. The Sikh also noticed a small cross on a chain round her neck. All the signs were that the unfortunate girl was dead. Just as he was about to switch off his torch, the officer glimpsed a trembling of the corner of her mouth. Was he mistaken? He knelt down, cleared one ear of the folds of his turban and pressed it to the young woman's chest, but her heart seemed to have stopped beating. Just in case, however, he called for a stretcher.

'Hamidia Hospital, quickly!' he shouted to the driver.

*

Hamidia! After Mahmoud Parvez's staff had returned with his wedding *shamianas*, the approaches to the great hospital looked like the encampment of some tribe struck down by a curse from above. In each tent, Parvez, who had recovered from gas inhalation, had unrolled mats, and set up tables and benches towards which the medical college students tried to channel the hordes of dying people who kept on pouring in. Picking out from this tide those who would benefit from a few blasts of oxygen or a cardiac massage was impossible. The white-smocked student who felt for Padmini's pulse was quite sure that the young woman brought in by Colonel Khanuja's soldiers was a hopeless case. As in wartime, it was better to work on those who had some chance of pulling through. He had her stretcher taken to the morgue where hundreds of corpses were already piled up.

In addition to pulmonary and gastric attacks, most arrivals were suffering from serious ocular lesions: burnt corneas, burst crystalline lenses, paralysis of the optic nerve, collapsed pupils. A few drops of atropine and a cotton wool pad for each eye was all the medical teams could offer their tortured patients. Seeing the cohorts of blind people stumbling over the bodies of the dying, Professor Mishra said to himself, 'Tonight the Bhopalis are going through their Hiroshima.'

⁂

Forty-eight-year-old Commissioner Ranjit Singh was the highest civil authority for the city of Bhopal and its region. As soon as he heard about the catastrophe, he jumped in his car and sped to the police headquarters in the heart of the old town. It was from this nerve centre that he intended to mobilize evacuation and rescue operations. Ranjit Singh would never forget his first glimpse of that hellish night. On the bridge running along the Lower Lake, he saw 'Tens, hundreds, thousands of sandals and shoes lost by people running away in their scramble to escape death.'

The Commissioner found the police headquarters in total disarray: gas had infiltrated the old building, burning the eyes and lungs of many of the officers. Yet calls were coming in, one after another without interruption, in the command room on the second floor. One of them was from Arjun Singh, Chief Minister of Madhya Pradesh. Rumour had it that he had left his official residence and taken refuge outside the city. Arjun Singh was calling in by radio to speak to the police chief Swaraj Puri.

'You must stop people leaving,' the head of the government said to him. 'Put barricades across all roads leading out of the city and make people go back to their homes.'

The Chief Minister, it seemed, had no idea of the chaos prevailing in Bhopal that night. In any case Puri had a good rebuttal.

'Sir,' he answered, 'how can I stop people leaving when

my own policemen have disappeared along with the other fugitives?'

The Commissioner decided to speak to the head of the government himself. He took over the microphone.

'Mr Chief Minister, no one can stop the human tidal wave trying to escape the blanket of gas. It's every man for himself. What's more, in the name of what do you want to stop these poor people from trying to save their lives?'

The senior official was suspicious of Singh's motives for wanting to stop the exodus. With one month to go to the general election, it was conceivable that the Chief Minister of Madhya Pradesh was afraid of losing votes. Had he not already taken the precaution of making sure of their support by handing out property deeds that legalized their squats beside the high-risk factory? This had been a decision the Commissioner had tried in vain to oppose for reasons of safety, but especially because it encouraged random settlement, the nightmare of any municipal authority. And now, when tragedy was striking the beneficiaries of Singh's largesse, the Chief Minister wanted to keep survivors in their homes. Indignant, the Commissioner curtailed their conversation and called his collectors to ask them to send all available vehicles to help evacuate the areas affected by the toxic cloud that was still spreading through a whole section of the city. Then, putting a damp towel over his face, he started up his Ambassador to drive through the columns of desperate fugitives, to the factory.

The spectacle he encountered at the entrance to the erstwhile pride of Bhopal was terrifying. Hundreds of people from districts to the north and east were banging on the doors of the dispensary where Dr Loya, Carbide's appointed doctor, and three overstretched nurses were trying to give a few breaths of oxygen to those most affected. On one of the four beds, with his face protected by a mask, lay the

329

only victim of the catastrophe on the factory's staff. Shekil Qureshi, who had believed as deeply in Carbide as he did in Allah, had been found sprawled at the foot of the boundary wall over which he had leapt after tank 610 exploded. The Commissioner was immediately conducted to the office in which Jagannathan Mukund had shut himself away. The first thing that caught his eye was a framed certificate on the wall, an award congratulating Mukund on his factory's excellent safety standards. 'But that night,' the Commissioner recounted, 'the recipient of that diploma was just a haggard man, annihilated by the magnitude of the disaster and by fear of a popular uprising.' Ranjit Singh tried to reassure him:

'I'll have armed guards posted at the entrance to the factory, as well as outside your residence.'

Suddenly, however, the Commissioner could no longer contain one burning question. 'I really wanted to know whether, for years, without my being aware of it, a plant located less than two miles from the centre of my capital, had been producing a pesticide, made out of one of the most dangerous substances in the whole of the chemical industry,' he would recall.

He recalled having read that in the United States people were put to death using cyanide gas.

'Did the gas that escaped from your plant tonight contain cyanide?' he asked.

Jagannathan Mukund grimaced before revealing what Carbide had always kept secret:

'In the context of a reaction at very high temperature, MIC can, in fact, break down into several gases, among them hydrocyanide acid.'

❈

All that night people called out for each other and searched for one another: in the Hamidia Hospital wards packed with bodies, in the streets, in the courtyard of the Jami Masjid, which its muftis had turned into an emergency reception centre. That night the water in the ablution tanks, diverted in days gone by from the Upper Lake by a British engineer, was a godsend. Victims rinsed their burning eyes and drank deeply in order to purge themselves of deadly molecules.

The tailor, Ahmed Bassi, the bicycle repairman, Salar, the worker-poet Rehman Khan availed themselves of the healing waters. Then they set off together in search of their families scattered by the disaster. In Spices Square, strewn with the bodies of poetry lovers and of hundreds of pigeons and parrots, Bassi and Salar met Ganga Ram carrying Dalima in her festival sari. After escaping the gun fire of the owner of the house in which they had sought refuge, the former leper and his wife had been miraculously spared by the gases and had headed directly south towards the mosque rather than towards the station. Such reunions illuminated an otherwise devastating night.

✼

In all this turmoil of suffering, fear and death, Sister Felicity did her best to save abandoned children in the corridors and wards of the hospital. There were dozens of them wandering about, almost blind, or lying groaning in their own vomit on the bare floor. The first thing the nun did was regroup them at the far end of the ground floor of the building where she had set up her help centre. Word got around quickly and other children were brought to her. Most of them had got lost during the night when their panic-stricken parents entrusted them to passengers in some truck or car.

With the help of two students, the nun carefully cleaned

their eyes that had been attacked by the vapours. Sometimes the effect was instantaneous. Her own eyes filled with tears when one of her protégés started to cry out: 'I can see!' She would guide those who had been miraculously cured to the aid centre and give her attention to other young victims, whom she bombarded with questions:

'Do you know this little girl?'

'Yes, she's my sister,' answered one child.

'And this boy?'

'He goes to my school,' answered another.

'What's his name?'

'Arvind,' a third told her.

Thus, little by little, the links between these devastated people were reestablished, and sometimes a distressed father or mother was reunited with a much-loved child.

A tall young man dressed in a festive *sherwani*, his feet shod in spangled mules, paced ceaselessly through the corridors and wards of that same hospital. He was looking for someone. Sometimes he would stop and gently turn a body over to look at a face. Dilip was sure that he would find Padmini somewhere in this charnel-house. He did not know that his young wife had just been carried away to the morgue on a stretcher.

The pot-bellied little man who had promised the police chief that if necessary he was prepared 'to feed the whole city' could never have imagined that he would have to keep his promise so soon. Shyam Babu, the proprietor of the Agarwal Poori Bhandar, the most famous restaurant in Bhopal, had just got to bed, when two men rang his doorbell. He recognized the president and the secretary of the Vishram Ghat Trust, a Hindu charitable organization of which he was a founding member.

'There's been an accident at Carbide,' announced the president before being overtaken by a coughing fit that sent him reeling. His companion continued.

'Thousands of people have been killed,' he declared. 'But, more importantly, there are thousands of injured who have nothing to drink or eat at Hamidia Hospital and under Parvez's *shamianas*. You, and you alone, can come to their aid.'

Shyam Babu stroked his moustache. His blue eyes lit up. May the goddess Lakshmi be blessed. At last he was going to fulfil his life long dream of feeding the whole city.

'How many are there of them?' he asked.

The president tried to overcome his bout of coughing:

'Twenty thousand, thirty thousand, fifty thousand, maybe more . . .'

Shyam stood to attention.

'You can count on me, no matter how many there are.'

As soon as his visitors had gone, he mobilized all his employees and enlisted the support of the staff of several other restaurants. Even before daybreak, some fifty cooks, assistants and bakers were at work making up rations of potatoes, rice, dal, curry and chapattis, which they wrapped in newspaper. Stacked into Babu's Land Rover, these makeshift meals were taken at once and distributed to the survivors. This was not to be the only good deed done by the restaurateur. Having taken care of the living, Shyam Babu would have to see to the dead.

Under the great tamarind tree in Kamla Park, the narrow strip of garden separating the Upper Lake from the Lower Lake, a *sadhu* watched impassively as people fled the deadly cloud. All through that night of panic, the Naga Baba, or naked holy man, as the Bhopalis called him, remained cross-legged in the lotus position. For thirty-five years he had lived there, ever since a five-day *samadhi*, a spiritual exercise in which he was buried alive, had turned him into a holy man. Half-naked, his body was covered in ashes and his long mop of hair divided into a hundred tresses. A pilgrim's stick topped with Shiva's trident and a bowl in which he collected food provided by the faithful were his only possessions. The Naga Baba, detached from all desires, material things, appearances and aversions, spent his days meditating, in quest of the Absolute. With prayer beads in his fingers, and his gaze seemingly vacant behind his half-closed lids, he seemed indifferent to the chaos that surrounded him. Overtaken by small, eye-level pockets of monomethylamine and phosgene borne along by the breeze, dozens of men and women with their lungs dilated from running, collapsed, asphyxiated around him. Trained to breathe only once every three or four minutes by his ascetic exercises, the Naga Baba did not inhale the vapours from the passing cloud. He was the only person to survive in Kamla Park.

# 43

## *The dancing girl was not dead*

The dead were everywhere. In the corridors, in the doctors' rooms, in the operating theatres, in the general wards, even in the kitchens and the nurses' canteen. Laid out on stretchers or on the bare floor, some looked as if they were sleeping peacefully; others had faces deformed by suffering. Strangely, they gave off no smell of decomposition, as if the MIC had sterilized anything in them that might rot. Removing these corpses became as pressing a problem as caring for the living. Already the vultures had arrived. Not the carrion birds, but the professional body riflers for whom the catastrophe was a godsend. Dr Mohammed Sheikh, one of the two doctors on duty, surprised one of these pillagers with a pair of pincers in his hands, preparing to yank gold teeth from the mouths of the dead. One of his accomplices was stripping the women of their jewels including their nose rings. Another was recovering watches. Their harvest was likely to be a thin one, however: Carbide's gases had primarily killed the poor.

Made aware of the problem, Professor Mishra sent some students to stand guard over the corpses and telephoned the two forensic pathologists at the medical college. The collector of vintage cars, Heeresh Chandra, and his young colleague who loved roses, Ashu Satpathy, were already on their way to the hospital. Chandra knew that the autopsies he and Satpathy would perform that night could save thousands of lives: the bodies of the dead could yield definitive information about the nature of the killer gases and might enable them to find an antidote.

What the two doctors saw on their arrival chilled them to the bone. 'We were used to death, but not to suffering,' Satpathy would recall. The hundreds of bodies they had to step over to gain access to the medical college looked as if they'd been tortured. 'What chemical substances could be capable of doing that kind of damage?' wondered Chandra who, before doing anything else, went rushing to the faculty library. His colleague Mishra had mentioned methyl isocyanate. The pathologist leafed frenetically through a toxicology textbook. The entry on the molecule did not contain much information, but Chandra suspected that it was capable of breaking down into highly toxic substances like hydrocyanide acid. Only hydrocyanide acid would be likely to inflict such deadly stigmata. As for Dr Satpathy, he went first to the terraces, to make sure that his roses had not been damaged by the toxic cloud. After examining every pot, plant, leaf and bud with all the concern and tenderness of a lover at his endangered mistress's bedside, he heaved a sigh of relief. The 'Black Diamonds' and 'Golden Chryslers' he had so lovingly grafted appeared to have survived the passing of the deadly fog. In two days' time, Satpathy would be able to show them, as planned, at the Bhopal Flower Show. Before returning to the inferno on the ground floor, he telephoned the third member of his pathology team, the photographer Subhash Godane.

'Get over here quickly, and bring a whole suitcase full of

film with you. You're going to have hundreds of photos to take.'

The young man who had dreamt of making his name photographing glamorously dressed women, hurriedly threw on his clothes, loaded his Pentax and hopped on his scooter.

Before performing their autopsies, the two pathologists had to carry out the essential task of setting up a system for identifying the victims. Nearly all of them had been caught in their sleep and had fled half naked. Satpathy enlisted the help of a squad of medical college students.

'Examine every corpse,' he told them, 'and jot down its description in a notebook. For example: circumcised male, approximately forty, scar on chin, striped underpants. Or again: little girl aged about ten, three metal bracelets on right wrist etc. Make a note of any deformities, tattoos and any distinctive features likely to facilitate identification of the victim by their next-of-kin. Then place a card with a number on it on each body.'

The doctor turned then to Godane.

'You photograph the numbered bodies. As soon as you've developed your negatives, we'll put them on display. In that way families will be able to try and find anyone they've lost.'

Next, addressing himself to everyone, he added:

'Hurry up! They'll be coming for the bodies soon!'

Soon the shutter release on the Pentax was firing like a tommy-gun over the stiffened bodies. Although he'd spent years immortalizing the victims of minor accidents on glossy paper, Subhash Godane was suddenly face to face with a wholly different form of death: industrial death, death on a huge scale. While he was working, he found himself wondering whether he had not photographed a particular attractive young woman in a multi-coloured sari, or a particular little girl whose long plaits had yellow marigolds in them, on a previous occasion. Perhaps on Hamidia Road, in the jewellery market at the great mosque, or near the fountain in Spices Square. But that night his models' eyes had rolled back, the amber tint of their skin had turned the colour of ashes, and their mouths had set into a dreadful rictus. Godane had difficulty continuing with his macabre documentary. All at once he thought he was seeing things. By the light of his flash, he saw the features of a face twitch. Two eyes opened. 'This man isn't dead!' he yelled to Dr Satpathy, who came running with his stethoscope. Sure enough, the man was still alive. The doctor called for a stretcher and had him taken to a recovery ward where he regained consciousness. He was wearing a railway worker's tunic. It was V. K. Sherma, the deputy stationmaster who had saved hundreds of passengers by risking his life to get the Gorakhpur Express to leave.

There were other shocks in store on that tragic night. Two female corpses were brought in by unknown persons. When Satpathy examined them, he realized that they had not been killed by gas but murdered. One had a deep wound to the throat, the other had burns to a substantial part of her body. The catastrophe had provided the killers with the perfect alibi. The doctor was also to see the corpse of the same little boy three times, labelled with three different numbers. It was a fraudulent act that would enable his family to claim three times the insurance compensation they believed the American multinational might pay.

Other parents refused to accept the awful reality. A young father placed his son's corpse in the arms of Dr Deepak Gandhe, one of the duty doctors.

'Save him!' pleaded the stranger.

'Your child is dead!' replied Gandhe, trying to give the little body back to his father.

'No! No! You can save him!'

'He's dead, I tell you!' insisted the doctor. 'There's nothing I can do for him.'

'Then the man ran off, leaving the child in my arms,' Deepak Gandhe recounted. 'In his heart of hearts he was convinced that I could bring him back to life.'

※

On dissecting the first corpses, the two pathologists could hardly believe what they found. The blood of a Muslim with a grey goatee beard, into which Satpathy dipped his finger, was as viscous as currant jelly. His lungs were ash-coloured, and a multitude of little bluish-red lesions appeared in a greyish frothy liquid. The man must have died drowning in his own secretions. Hearts, livers and spleens had tripled in size, windpipes were full of purulent clots. All the organs without exception seemed to have been ravaged by the gas, including the brains, which were covered with a gelatinous, opalescent film. The extent of the damage was terrifying even to specialists as hardened as old Chandra and his young colleague. A smell confirmed their suspicions as to the nature of the agent responsible – a smell that was unmistakable. All the corpses on which they performed autopsies gave off the same smell of bitter almonds, the smell of hydrocyanide acid. Here was the confirmation of what Jagannathan Mukund had let slip to Bhopal's commissioner. When it broke down, MIC released hydrocyanide acid, which instantly destroyed the cells' ability to transport oxygen. It was hydrocyanide acid that had killed the greatest majority of Bhopalis who died that infernal night. The two pathologists' discovery was vitally important, because hydrocyanide acid poisoning had an antidote: a commonplace substance, sodium thiosulphate or hyposulphite, well known to photographers who use it to fix their negatives. Massive injections of hyposulphite might possibly save thousands of victims. Chandra and Satpathy rushed to Professor Mishra, who was orchestrating the medical aid with his team. Strangely, the professor refused to believe his colleagues' findings and follow their recommendations. As far as he was concerned, the presence of hydrocyanide acid was an invention of the pathologists' over-fertile imaginations.

'You take care of the dead and let me take care of the living!' he directed them.

No one would really be able to account for this reaction on the part of the illustrious professor. It would deprive the afflicted of treatment that might have saved their lives.

Dawn broke at last on that apocalyptic night: a crystal clear dawn. The minarets, cupolas and palaces were lit up by the sun's thousand rays and life asserted itself once more in the entanglement of alleyways in the old part of town. Everything seemed the same. And yet some places looked like ossuaries on the morning after a battle. Hundreds of corpses of men, women and children, cows, buffaloes, dogs and goats were all over the place. Deeply alarmed by this state of affairs, Commissioner Ranjit Singh went to nearby colleges in areas that had been spared and enlisted students to pick up bodies. At the Maulana Azad technical college, he found dozens of volunteers.

'Divide yourselves up into two teams,' he told them. 'Muslims in one, Hindus in the other, and each can look after their own dead.'

His suggestion provoked a vehement reaction.

'Is there any difference between Hindus and Muslims at a tragic time like this?' objected one student.

'Is there even a god when such a catastrophe is allowed to happen?' said another.

'I made myself very small,' the Commissioner recalled, 'I was trying to think of the strongest possible terms in which to thank them.'

With bandannas over their mouths and noses, the students set off on scooters for the slums that Colonel Khanuja and his trucks had partially evacuated during the night. There were still a few survivors left among the mass of bodies. Student Santosh Katiyan was party to a scene that moved him deeply. While he was preparing to remove the body of a Muslim woman from one of the huts in Chola, a hand stopped him. A woman he recognized by the red dot on her forehead as a Hindu, slipped all her bracelets off her wrist and slid them onto her dead neighbour's arm.

'She was my friend,' she explained, 'she must look beautiful to meet her god.'

A little further on, Santosh noticed four veiled Muslim women, sitting under the porch roof to a small Hindu temple. They were consoling a woman who had lost her entire family. In such extreme distress, distinctions of religion, caste or background vanished. Very swiftly, however, the sordid took its place alongside the sublime. No sooner had Rajiv Gandhi announced over the radio that all families would be compensated for the loss of their loved ones, than people began to squabble over the corpses. Outside the medical college, Colonel Khanuja saw two women pulling the body of a man by his hands and legs in opposite directions. One was a Hindu; the other Muslim. Both were claiming that the deceased was her relative. They were pulling so hard that the poor man was in danger of being torn in two. The Major decided to intervene.

'Undress him! Then you'll see whether or not he's circumcised.'

The two women tore off his *lunghi* and underpants and examined his penis. The man was circumcised. Furious, the Hindu woman got up and set off in search of another corpse.

❋

The number of expressions of solidarity multiplied. Never before had the India of 1000 castes and 20 million divinities shown itself so united in adversity. Tens of organizations, institutions, associations, hundreds of entrepreneurs and businessmen, thousands of private individuals of all social classes, the Rotarians, the Lions, the Kiwanis, and the scouts, all came rushing to the rescue of the survivors. Many towns in Madhya Pradesh sent truckloads of medicines, blankets and clothing. Volunteers of different religious faiths spread out cloths on the corners of avenues, in squares, all over the place, onto which people threw mountains of rupees.

That day after the catastrophe was also a time for anger. A policeman came to warn Mukund, who had remained closeted in his office, that thousands of rioters were heading for the factory, yelling: 'Death to Carbide!' After trying all night to get hold of his superiors in Bombay, the engineer finally got through by telephone to one of them.

'There's been an accident,' he informed his boss, K. S. Kamdar. 'An MIC leak. I don't know yet how or why.'

'Any fatalities?' Kamdar asked anxiously.

'Yes.'

'Many?'

'Alas! Yes.'

'Two figures?'

'More.'

'Three?'

'More like four, Kamdar.'

342

There was a long silence at the other end of the line. Kamdar was stunned. At last he enquired:

'Do you have the situation in hand?'

'Until the crowd invades the factory. Or the police come and arrest me.'

At that point their conversation was interrupted by several uniformed policemen and two plain-clothed inspectors from the Bureau of Investigation. They carried a criminal warrant to detain Mukund and his assistants.

Outside the situation was worsening. Swaraj Puri, the police chief who had seen so many of his men disappear the previous night, feared violent action. With no means of opposing it, he decided to resort to a stratagem. He summoned the driver of the only vehicle left to him with a loudspeaker.

'Drive all over town,' he ordered him, 'and announce that there's been another gas leak at Carbide.'

The effect of the ruse was miraculous. The rioters who had been about to overrun the factory scattered instantly. In a matter of minutes the city was empty. Only the dead remained.

*

The fatal cloud had spared the vast enclosure at the end of Hamidia Avenue where, in the shade of century-old mango and tamarind trees, generations of Muslims had been laid to rest. The man in charge of the place was a frail little individual with dark skin and a chin studded with a small salt and pepper goatee. Abdul Hamid had been born in that cemetery. He had grown up there and become its master. It was a position which enabled him to live in comfort: for every burial he received 100 rupees and he oversaw two or three every day. Abdul Hamid was a key figure in the Muslim community. They all, at one time or another, had dealings with him. He was no stranger to death, but the poor man could never have anticipated the spectacle that awaited him that morning at the entrance to his cemetery. Dozens of bodies wrapped in shrouds were piled up like parcels outside the grille. 'It was the first time I'd ever seen so many corpses at once,' he later recalled.

Hamid called his sons and set about digging graves. Volunteers came to help him. But how was he to give so many dead a decent burial? How was he to receive their families appropriately? In the absence of any members of the clergy, it was Abdul himself or one of his gravediggers who recited the prayer for the dead. In a few hours there was nowhere left to dig fresh holes; the men had to stop for fear of disturbing the remains of earlier burials. 'I was the guardian of the dead,' Abdul Hamid would say later. 'I had no right to violate tombs. If I did, no one would trust me any more.'

In the two other Muslim cemeteries, the congestion was even worse, a fact that forced the city's Grand Mufti, the venerable Kazi Wazid ul-Hussein, to issue an urgent fatwa authorizing the disturbance of old tombs in order to make room for Carbide's victims. The fatwa stipulated that some ten bodies could be buried in the same grave. Soon a flood of trucks, cars and handcarts turned up with their macabre loads. The deceased were deposited at the entrance to Abdul Hamid's cemetery in the columned building set aside for preparation of the dead. In the absence of relatives, this ritual was carried out by volunteers, who undressed the bodies and washed them in tepid water. Men and women were dealt with separately. The elderly Iftekar Begum, the 80-year-old dowager who directed operations, marvelled that so many of the deceased were wearing embroidered *burkahs* and flowers in their hair.

'Last night was Sunday,' a friend explained to her, 'they died while they were celebrating.'

Other surprises awaited those dealing with the burial of the dead. Under pressure from the gases produced by the chemical decomposition of MIC, the corpses were subject to strange twitches. Here an arm stretched itself out, there a leg. Some bodies buried near the surface seemed to want to stand up. Terrified by these extraordinary 'resurrections', some people fainted, others shouted at the ghostly apparitions, yet others ran away, screaming. Abdul Hamid was struck dumb with bewilderment: his cemetery had become a theatre of ghosts.

⁂

Bhopal's most celebrated restaurateur had been obliged to hand over his ovens to his two sons and two sisters while he arranged for the Hindu funeral pyres. His associates from the Vishram Ghat Trust, the group in charge of cremations, were overwhelmed by the magnitude of the task. The Hindu religion ordains that, with the exception of children, the bodies of the deceased must be burnt. For that they needed firewood, but how were they to find enough for thousands of corpses? Shyam Babu worked a miracle. In the space of a few hours, he managed to fill fifteen trucks with enough wood to incinerate several hundred bodies. Cloth makers brought him miles of linen with which to make shrouds.

While he prepared to set light to the first pyre, two envoys of the Mufti appeared. They had come to make certain that no Muslims would be burned by mistake. It was almost impossible to confuse men from the two communities: the adherants of Islam wore a characteristic goatee, amulets round their neck and bore marks left on their foreheads by their repeated prostrations. Not to mention the fact that they were circumcised. Unless they were veiled with their *burkahs*, women were more difficult to distinguish. Nonetheless, the Mufti's envoys left reassured. Shyam Babu was just about to plunge his torch into the pile of wood when someone grabbed his arm. The student Piyush Chawla had spotted a little gold cross round one young woman's neck.

'This woman isn't a Hindu!' he cried and promptly extricated the body and placed it to one side of the pyre.

Then he noticed an almost imperceptible quivering of her eyelids. Intrigued, he bent over the body. The hands and feet were neither rigid nor cold. This woman with bells on her ankles was not dead, he was sure of it. He put her on one of the trucks that was going to bring back other corpses from Hamidia Hospital and climbed up beside her. Frothy bubbles were coming out of her half-open mouth. Piyush Chawla could not help wondering whether he was witnessing some supernatural phenomenon.

It was exactly two in the afternoon by the clock in Spices Square on that Monday 3 December when the smoke from the first funeral pyre rose into the sky over Bhopal, reducing to ashes those people to whom Carbide's 'beautiful plant' had promised happiness and prosperity. Blowing now from the south, a light breeze carried away the last traces of deadly gas and replaced them with a smell even more appalling: the stench of burnt flesh.

# 44

## 'Death to the killer, Anderson!'

Tuesday 4 December, 8.30 hours. The athletic figure of the CEO of Union Carbide made his entrance into the boardroom at the company headquarters in Danbury. Since the previous day, Warren Anderson had been given hourly reports on the situation in Bhopal. For a son of immigrants who had managed to haul himself up to the top of the world's third largest chemical giant, the tragedy was as much a personal disaster as a professional one. Anderson had set his sights on making Union Carbide an enterprise with a human face. Of the 700 industrial plants he ran, employing 117,000 people in thirty-eight countries, the factory built in the heartland of the Indian subcontinent had been his favourite. It was he who had inaugurated it on 4 May 1980. The first drops of MIC that emerged from its distillation columns that day had been his victory. Thanks to the Sevin thus produced, tens of thousands of Indian peasants would be able to conquer the menace of famine.

As soon as he heard about the tragedy, Anderson had set up a special team to deal with events in total transparency. He had arranged for the media to maintain a constant press link to the company spokespeople. Then he had shut himself away in his office in his home in Greenwich to think about what his initial reaction should be. Having made his decision, he informed his closest colleagues at once by telephone. Despite the terrified entreaties of his wife Lilian, he would leave immediately for Bhopal. His place was right there, among the victims. He wanted to see for himself that everything that could be done was being done. His gesture would help underline the fact that the company he controlled was not a faceless, soulless giant, and that the tragedy which had just occurred was just one accident along a path intended to create a better, more just world. In short, his presence at the scene of the catastrophe would be an expression of the ideal that inspired him.

In addition to a sense of moral obligation towards the victims, he also felt a responsibility to the company's shareholders. Doubtless Carbide had the financial means to survive the worst possible disaster. But if the terrible news he had received was accurate, his duty was to do everything in his power to prevent the company from appearing cruel or irresponsible to its shareholders.

By the sombre faces that greeted him that Tuesday morning in the presidential boardroom at Danbury, Warren Anderson could tell that his colleagues were hostile to his idea. There was no shortage of arguments against it. Firstly, he would be risking his life. India was an unpredictable country. One month earlier, Indira Gandhi had been assassinated after her army had killed far fewer people than had died at Bhopal. Some survivor, crazed by the loss of a loved one, might just make an attempt on his life. Then again, under pressure from outraged public opinion, the Indian government might imprison him on arrival. Either way, his journey risked giving the unnecessary impression that the multinational was directly responsible for the tragedy, when it would be better to let its Indian subsidiary take all the blame. There was also the fact that there was every likelihood that the visit would be perceived as a provocation. Finally, it would expose the company Chairman to dangerous confrontations with India's new political authorities, with the Press, lawyers, judges, diplomats . . . Even those in charge of the Indian subsidiary who had been consulted over the telephone showed little enthusiasm for the idea of having their top man arrive at the scene of the accident. Anderson, however, had made up his mind.

'I've weighed all the risks,' he declared, 'and I'm going.'

❋

On Thursday 6 December at five o'clock in the morning, a Gulf Stream II twin-jet plane landed at Bombay's Santa Cruz airport. No one took any notice of the three initials engraved on its crest: yet they belonged to the American company that had just inflicted death upon the country. Suffering from the flu, exhausted after the twenty-hour flight, Warren Anderson travelled discreetly to the luxurious Hotel Taj Mahal opposite the symbolic arch of the Gateway of India, where a suite had been reserved for him. The two Indian gentlemen there to welcome him, Keshub Mahindra, President of Union Carbide India Limited, and V. P. Gokhale, its Managing Director, brought him up to date with the latest figures from the accident. By then people were talking about 3000 dead and 200,000 people affected. Fortunately, the two Indians also had some good news: Arjun Singh, the Chief Minister of Madhya Pradesh, and Rajiv Gandhi, head of the country's government, had agreed to see Carbide's Chairman. That was one source of satisfaction for Anderson: he could at least convince them of his company's readiness to compensate the victims, starting with at least 5 million dollars' worth of emergency medical aid.

Not wanting to draw attention to themselves, Anderson and his two partners flew to Bhopal the next day in a Boeing 737 on a regular Indian Airlines flight. The company jet would rejoin the Chairman in New Delhi to take him back to the United States.

On landing, the American noticed a small group of policemen on the tarmac. 'How tactful of the local authorities to have sent us an escort,' he thought. As soon as the steps were in position, two officers climbed aboard and a voice came over the cabin address system: 'Mr Anderson, Mr Mahindra and Mr Gokhale are invited to leave the aircraft first.'

Ah, the wonders of Indian hospitality! Police chief Swaraj Puri, who on the night of the tragedy had watched his policemen flee, was at the foot of the plane in the company of the city's Collector to welcome the visitors with warm handshakes. All that was missing was the traditional garland of flowers and a pretty hostess to give them a welcoming *tilak*. Anderson and his companions took their seats in an official Ambassador brought to the foot of the steps. The car took off like the wind and left the airport via a service gate to avoid the pack of journalists waiting in the arrivals hall. The police chief and the Collector followed in a second car.

'Thank you for having gone to the trouble of fetching us,' Anderson said to the uniformed inspector sitting beside the driver.

'It's standard procedure, Sir. There's considerable tension in the city. It's our duty to look after your safety.'

The American took pleasure in being back in the city, the beauty of which he had so admired at the time of the factory's inauguration four years earlier. The minarets of the mosques casting their reflections in the waters of the lake, the numerous parks brimming over with flowers, the picturesque old streets bustling with activity; everything seemed so normal that he found it difficult to believe that the city had just been through so dreadful a nightmare.

The car climbed towards Shamla Hill, entered the grounds of the Carbide research centre and stopped in front of the company's splendid guest house. Anderson was astonished to find two squads of policemen assembled on either side of the door to the establishment. An officer was waiting on the steps. As soon as the three visitors got out of the car, he stepped forward, came to attention and saluted. Then he announced:

'I regret to inform you that you are all three under arrest.'

Anderson and his partners started with surprise. The policeman continued: 'Of course, this is a measure primarily for your own protection. You are free to come and go about your rooms, but not to go out or use the telephone, nor to receive visitors.'

At that moment the police chief and the Collector arrived. They were accompanied by a magistrate in his distinctive black robe. The American felt reassured: there had been some misunderstanding. They were coming to set them free. In fact, the magistrate had been summoned to notify the visitors of the reasons for their arrest. He informed them that by virtue of articles 92, 120B, 278, 304, 426 and 429 of the Indian penal code, they were accused of 'culpable homicide causing death by negligence, making the atmosphere noxious to health, negligent conduct with respect to poisonous substances, and mischief in the killing of livestock.' The first charge was punishable with life imprisonment, the others carried sentences of between three and six months.

'Naturally, all those charges carry the right to bail,' intervened Keshub Mahindra, President of Carbide's Indian subsidiary.

'I'm afraid that is, unfortunately, not the case,' replied the magistrate.

'What about our meeting with Chief Minister Arjun Singh?' asked the American anxiously.

'You will be notified about that as soon as possible,' the police chief informed him.

⁂

The likely instigator of this brutal reception was absent from Bhopal. He had left the capital of Madhya Pradesh that very morning to join Rajiv Gandhi on an electoral tour. He had, however, left instructions with his spokesman. As soon as the three visitors had been arrested, the latter was to muster the press and deliver the news with maximum impact. Arjun Singh, though a long-standing friend of Carbide, expected to make the most of his audacity. By having the American company's Chairman and his Indian partners arrested, he was setting himself up as the avenger of the catastrophe's victims, a move that could only help him in the next parliamentary election. 'The government of Madhya Pradesh could not stand passively by and watch the tragedy,' his spokesman told journalists on his boss's behalf. 'It knows its duty to the thousands of citizens whose lives have been devastated by the criminal negligence of Union Carbide's directors.'

News of Warren Anderson's arrest created a sensation from one end of the planet to another. This was the first time that a Third World country had dared to imprison one of the West's most powerful industrial leaders, even if his prison was a five star guest house. In New Delhi there was consternation. The Indian Foreign Affairs Minister had promised the American State Department that nothing would impede Anderson's journey. Quite apart from wishing to avoid an overt clash with the United States, Indian leaders were afraid that the incident would dissuade

352

large foreign firms from setting up in India for ever. The Chief Minister of Madhya Pradesh would have to release his prisoners immediately. Never mind justice. Matters of state required it.

⁂

Three hours later, Bhopal's chief of police assisted by several inspectors came to announce the release of the American prisoner. His Indian colleagues would be set free some time later.

'A government airplane is waiting to take you to Delhi, from where you will be able to return to the United States,' he stated precisely.

He then presented him with a document. To his stupefaction Anderson discovered that the sum of 25,000 rupees, about 2000 dollars at the time, had been paid by his company's local office as bail. He had only to declare his civil status and give his signature and he would be free. 'Twenty-five thousand rupees for the release of the head of a multinational responsible for the deaths of three thousand innocent people and the poisoning of two hundred thousand others! What does that make an Indian life worth?' enquired the Indian Press the next day.

The news created an immediate uproar in the pack of reporters jostling with each other at the entrance to the guest house. The most significant reaction, however, came from a crowd of demonstrators pressed to the railings of the research centre compound. From the car bearing him away to the airport, Warren Anderson could see a forest of placards above their heads. The sight of the few words inscribed on the pieces of cardboard would haunt him for the remainder of his days. 'Death to the killer, Anderson!' chanted the people of Bhopal.

The Chairman of Union Carbide would never meet Prime Minister Rajiv Gandhi or any of his ministers. Only an official in the Foreign Office would agree to give him a brief audience, provided the Press was not informed. The man who had hoped to change the living conditions of India's peasants and who had wanted, as he had stated, to retire 'in a blaze of glory', left India broken, humiliated and despondent. He still did not know exactly what had happened in his 'beautiful Indian plant' on the night that spanned the second and third of December. As for his desire to provide the victims with aid, he had not even been able to discuss it. His journey had been a fiasco. A few minutes before he climbed into his Gulf Stream II and took off for the United States, a journalist challenged him:

'Mr Anderson, are you prepared to come back to India to answer any legal charges?'

Anderson turned pale. Then in a steady voice, he replied:

'I will come back to India whenever the law requires it.'

In the meantime, other Americans had been landing in Bhopal. Danbury had rapidly dispatched a group of engineers whose mission it was to shed light on the catastrophe. Naturally the factory's last American Managing Director was part of that delegation. For Warren Woomer, this return was a painful trial. 'My wife Betty and I had spent two of the best years of our lives here. But now I'd come back to examine the remains of a factory which had in a sense been my baby,' the engineer would recall. He had real difficulty recognizing it. The ship he had left in good working order was now a spectacle of desolation that tore at his heart strings. He made an effort to stay calm during his first conversation with his successor Jagannathan Mukund. Why was there so much MIC stocked in the tanks? Why were all the safety systems deactivated? Woomer fumed silently. The enquiry team had, however, agreed that they would avoid any confrontation. The important thing was to gather as much information as possible, not to create controversy.

The task threatened to be impossible, however, because men from India's Criminal Bureau of Investigation had taken over the affair. Their chief, V. N. Shukla, was a stiff-necked, unsmiling man. He began by prohibiting the Americans access to the plant. Then he told Woomer:

'If I catch you, or any of your colleagues, interrogating any of the workmen, I'll throw you in prison.'

Worse yet the CBI was also in the process of moving the factory's archives to a secret location. What were the American investigators supposed to do, given that they could not examine the site, question witnesses or refer to such crucial documents as reports of procedures carried out on the fatal night? Woomer felt overwhelmed. Especially as the situation was further complicated by the arrival on the scene of a team of Indian investigators headed by a leading national scientist, Professor Vardarajan, president of the Indian Academy of Science. How could they cope with this competition and the police restrictions? Woomer soon passed from feeling overwhelmed to despair. Once again, however, the good fairy of chemistry came to the rescue of its disciples. One thing upon which they were all agreed was that before beginning an investigation, they needed to be certain that no further accidents could occur. It was this concern that haunted Woomer. There were still twenty tons of MIC in the second tank and one ton in the third. At any moment, those deadly substances could start to boil. On this, Americans and Indians were in accord. Should they repair the flare and burn the gases off at altitude? Should they decontaminate them with caustic soda after getting the scrubber back in working order? Should they try and decant them into drums and evacuate them to a safe place? In the end it was Woomer who came up with the solution.

'Listen,' he said, in his nonchalant but reassuring voice, 'the best way to get rid of the remaining MIC is to use it to make Sevin.'

'But how?' asked the Indian professor, stupefied.

'By getting the plant running,' replied Woomer. 'After all, that was what it was built for.'

Making Sevin meant cleaning all the pipework, pressurizing the tanks, repairing the faulty stopcocks and valves, reactivating the scrubber and the flare, lighting the alpha-naphtol reactor again . . . It meant re-engaging all the systems of a plant, the wreckage of which had just caused a catastrophe unprecedented in human history.

'How long would it take you to attempt such an operation?' asked the Indian professor.

'No more than five or six days,' answered Woomer.

'And what about the local people? How are they going to react when they hear the factory's going into operation again?'

The American engineer could not answer that question. Someone else was going to do it for him.

# 45

## 'Carbide has made us the centre of the world'

The Chief Minister of Madhya Pradesh was exultant. Warren Woomer's idea would enable him to wipe out the memory of his alleged absence on the night of the tragedy and win back his electorate. This time he would be seen right there on the battlefield. To ensure that his heroism paid off, he would need to convince the people of Bhopal that restarting the factory would be extremely dangerous. Arjun Singh promulgated several safety measures with no real purpose other than to create an atmosphere of panic. He ordered all schools closed, despite the fact that they were in the middle of exams and most were situated outside the risk zone. Next he called in 800 buses to evacuate all those living within a two and a half mile radius of the factory. Once people were well and truly terrified, he revealed the plan from which he would emerge a great man. He dispatched an army of motorized rickshaws equipped with loudspeakers across the city. The whole of Bhopal then heard his steady, reassuring voice: 'I have decided to be present in person in the Carbide factory on the day when its engineers start it running again to remove the last drops of any toxic substances,' he declared. 'This moment of truth will be a token of your humble servant's dedication to your cause. This is not an act of courage, but an act of faith and that is why I am calling this challenge to get rid of any residual dangers at the cursed factory, *Operation Faith*.'

As the fateful day for restarting the factory approached, businesses closed, streets emptied and life came to a halt. The Chief Minister let the exodus become a torrential flood. Driven by the fear that he had so adroitly stirred up, people threw themselves into his 800 buses and into any other means of transport. They abandoned their homes in buffalo carts, rickshaws, scooters, bicycles, trucks, cars and even on foot. The railway station was taken by storm. Afraid that their homes would be pillaged, people took with them anything they could. One woman left with her 9-month-old goat in her arms. For the oldest, the sight of trains covered with people piled on the roofs, and hanging from the doors and steps, brought back sinister memories of India's Partition. 'This spontaneous migration,' wrote the *Times of India*, 'defies all reason.'

✳

The newspaper was right. Bhopal had lost all reason. Yet, as Ganga Ram and Dalima were to find, to their astonishment, on their return to Orya Bustee, it was not in the place worst affected by the gases that the terror raged most intensely. If anything the reverse was true. Their neighbours might look like ghosts with their cotton wool pads on their eyes, but they were no longer afraid. Although the deaths of Belram Mukkadam, Rahul, Bablubhai, Ratna Nadar, and old Prema Bai and so many others had created an irreparable void in their small community, the joy of being reunited with friends was stronger than the fear of another disaster. The reunion of Ganga and Dalima with

Sheela and Gopal, Padmini's mother and brother; with Iqbal, Salar and Bassi, to name but a few, were occasions for celebration. What a joy it was to discover that Padmini was still alive in Hamidia Hospital and that Dilip was with her! What a relief to find one's hut intact when so many others had been looted!

Ganga and Dalima realized at once that the priority for people in their area was not to flee from a fresh threat but to preserve the fragile thread that attached them to the world of the living. Most had been seriously affected by the gases. They were in urgent need of medication. The hospital supplies had been exhausted, so costly treatment would have to be bought from pharmacies. But with what? Ganga would never forget the sight of his neighbours rushing to the only person now in a position to help them. Since the catastrophe, the money lender Pulpul Singh's house had been besieged by survivors clutching the title deeds for their hut, transistor radios, watches, jewels or anything else they had, in the hope of exchanging them for a few rupees. People jostled with each other outside the grille, threw themselves at the Sikh's feet, pleaded and paid him every conceivable compliment. As impassive as a Buddha, he made a clean sweep of all that he was offered. His wife and son noted down names, took a thumbprint on the receipts by way of signature, and arranged a most unusual array of objects all over their house. Even chickens that had survived the fateful night could bring in a few notes. That evening, a large box carefully wrapped in a blanket also found its way into the usurer's treasure trove: Ganga Ram had pawned his television set. With the money he received he would be able to help his neighbours get medicine to relieve their suffering. The magic screen that had brought his brothers and sisters in Orya Bustee so many dreams would have to wait for better days to foster other fantasies.

By 16 December, the day of *Operation Faith*, Bhopal was a ghost town, but television cameras were going to broadcast an event that would be larger than life. Since dawn, fire trucks had been spraying the streets to neutralize any suspect emanations. Over 5000 gas masks had been stored at the city's main crossroads. A cordon of ambulances and fire engines isolated the factory, while several hundred policemen posted at the various gates filtered through the holders of special permits. Among them were the Chief Minister and his wife. They would both be in the front line. Under the photographers' flashes they took their places in the control room, where Shekil Qureshi and his team had been on watch on the night of 2 December. Three military helicopters equipped with water tanks, their pilots wearing gas masks, circled continuously over the metal structures, ready to intervene should the need arise. 'To think that it took the death of thousands of people for our government finally to take an interest in our factory,' said one disillusioned workman as he followed the progress of the operation on his transistor.

Warren Woomer was satisfied: the equipment necessary to get things running again had been repaired in record time. At eight o'clock precisely, Jagannathan Mukund, surrounded by a police guard, was able to open the stopcock and allow hydrogen to flow into tank 611. A few minutes later, a supervisor announced that the tank had reached the correct pressure, which meant that they could start evacuating the first gallons of the twenty tons of MIC into the reactor, to make Sevin. At 13.00 hours, Professor Vardarajan let the Chief Minister know that one ton of methyl isocyanate had been turned into pesticide.

Arjun Singh was triumphant. *Operation Faith* had made a totally successful start. Draining the tanks to the last drop of MIC would take three days and three nights. Beaming happily, the intrepid politician clattered down the metal staircase of the 'beautiful plant' with his wife. Already his fellow citizens were preparing to return to their homes. Now he was sure of it: in two months' time they would turn out *en masse* to vote for him.

❄

'Everyone to the tea house! There's a *sahib* there who wants to talk to us!'

Since Rahul's death, young Sunil Kumar had taken over as messenger in the alleyways of Orya Bustee. He had lost all seven of his brothers and sisters, as well as his parents, in the catastrophe. They had all arrived recently from the countryside scorched by drought. The news he spread from hut to hut that morning brought a throng of survivors to the meeting place.

The 'ambulance chasers' had arrived. They had come from New York, Chicago and even California, people such as the celebrated and formidable San Francisco lawyer Melvin Belli, who announced that he was lodging a writ for compensation against Carbide for a mere 15 billion dollars, more than twice the amount of international aid India would receive that year.

The tragedy made fine pickings for that special breed of American lawyer who lives off other people's misfortunes and specializes in obtaining damages and compensation for the victims of accidents. The 400,000 or 500,000 Bhopalis affected by the multinational's disaster represented tens, possibly even hundreds of millions of dollars in various indemnities. Under American law, lawyers could collect almost a third of that sum in professional fees, a colossal bounty that transformed the office of Bhopal's mayor and that of the

Chief Minister into battlegrounds for vested interests. Like big game hunters, the Americans fought over clients and the various hunting grounds. The Kali Grounds fell to the representative of a New York law firm. Chaperoned by the godfather Omar Pasha, accompanied by an escort of Indian associates and two interpreters, 42-year-old lawyer Frank Davolta Jr, a half-bald colossus of a man, entered Orya Bustee in a swarm of policemen and press and television reporters. The escorts took up their position round the wobbly tea house tables. Servants brought baskets full of snacks, sweets and bottles of Campa Cola for the American to hand out. After the horror of the last few days, Orya Bustee was rediscovering its capacity for celebration.

When the first survivors appeared, the American had difficulty in repressing a feeling of nausea. Many of them were blind, others dragged themselves along on sticks or lay sprawled out on stretchers. They all gathered in a semicircle on the sisal mats that had been used for Padmini's wedding feast. The lawyer looked up with incredulity at the source of all this horror. In the winter sunshine, the Carbide plant stood glinting a few hundred yards away like one of Calder's mobiles.

Ganga Ram surveyed the *sahib* with suspicion. This was the first American ever to come into Orya Bustee. Why was he there? What did he want? Was he some envoy from Carbide come to convey the company's apologies? Was he the representative of some sect or religion wanting to say prayers for the dead and those who had survived? It would not be long before the survivors discovered the purpose of his visit.

The American lawyer stood up.

'Dear friends,' he said warmly, 'I've come from America to help you. The gas killed people who were dear to you. It ruined the health of those close to you for ever, possibly yours too.' He pointed to the factory on the other side of the parade ground. 'The Union Carbide company owes you reparation. If you agree to entrust the defence of your interests to me, I will fight for you to receive the highest possible compensation in my country's courts.'

The lawyer paused to allow his interpreters to translate his words into Hindi, then into Urdu and Orya. A man in a turban wagged his head, relishing every word. Not for anything in the world would Pulpul Singh have missed this event. He was already contriving ways of diverting this prospective manna into his safe.

Yet the American was surprised at what little reaction his proposal seemed to engender. The faces before him remained set, as if paralysed. Omar Pasha tried to reassure him: be patient, the gas damaged many of the survivors' mental faculties. This explanation further engaged the lawyer's interest. He decided to question some of the victims. He wanted them to tell him about the dreadful night, to describe the suffering to him. He invited everyone to talk about those they had lost. Sheela Nadar, Iqbal, Dalima and Ganga Ram spoke in turn. Suddenly the ice was broken. Calamity found a face and a voice. Frank Davolta took notes and photographs. He felt his file taking shape, assuming a life, gaining weight. Each testimony moved him a little more. By now, he was breathing so heavily that he had to undo his tie and open his collar. Moved to pity, Dalima came to his rescue. She brought him a glass of water, which the American gratefully emptied. He did not know that the water came from the well in Orya Bustee that had been contaminated by Carbide's waste with lead, mercury, copper and nickel, the water that the condemned of the Kali Grounds had been drinking for twelve years.

While the baskets of snacks were passed round, the lawyer resumed his speech:

'My friends,' he explained, 'if you agree to my representing your interests, we must draw up a contract.'

Upon these words, an assistant passed him a file of forms that he brandished at arm's length. 'These are powers of attorney,' he explained, 'authorizing counsel to act in lieu of his client.' The residents of Orya Bustee, who had never seen such documents, got up and thronged round the American's table. Like thousands of other Bhopalis from whom American lawyers extracted signatures that day, they could not make out the words printed on those sheets. They were content just to touch the paper respectfully. Then Ganga Ram's voice rose above the crowd. The former leper asked the question that was on everyone's lips.

'*Sahib*,' he asked, 'how much money will you be able to get for each of us?'

The lawyer's features froze. He paused as if thinking, then blurted out:

'No less than a million rupees!'

This unheard of figure struck the assembly dumb.

'A million rupees!' repeated Ganga Ram, unable to hold back his tears.

The television lenses closed in on him as if he were Shashi Kapoor, the star of the big screen. Cameras flashed.

'Are you surprised at the sum?' asked one reporter.

'No, not really,' stammered the former leper.

'Why not?' pressed the reporter.

Ganga pointed a fingerless hand at the pack of journalists jostling round him.

'Because Carbide has made us the centre of the world.'

# Epilogue

No one will ever know exactly how many people perished in the catastrophe. Concerned with limiting the amount of compensation that would eventually have to be handed out, the authorities stopped the reckoning quite arbitrarily at 1754 deaths. Reliable independent organizations recorded at least 8000 dead for the night of the accident and the following two days.

In fact, a very large number of victims were not accounted for. Among them were many immigrant workers with no fixed address. Sister Felicity and several survivors from the neighbourhoods on the Kali Grounds reported having seen army trucks on the morning of 3 December picking up piles of unidentified corpses and taking them away to some unknown destination. Over the next few days, numerous bodies were seen floating on the sacred River Narmada, whose sandy shores had helped to produce the first sacks of Sevin. Some of them drifted as far as the Arabian Sea, more than six hundred miles away; others fell prey to crocodiles.

In the absence of official death certificates, large numbers of corpses were incinerated or buried anonymously. Per the Mufti's order, the gravedigger Abdul Hamid found himself having to bury up to ten Muslims in the same grave. According to the restaurateur Shyam Babu, who supplied the wood for cremations, more than 7000 corpses were burnt on the Vishram Ghat Trust's five funeral pyres. The Cloth Merchant Association, for its part, stated that it had supplied enough material to make at least 10,000 shrouds for the Hindu victims alone.

The authorities contested the accuracy of these figures on the grounds that they exceeded the number of claims filed for compensation. This official reaction did not, however, take into account the fact that in many instances the tragedy had wiped out whole families and there was no one left to apply for damages. Four hundred dead, whose photographs remained posted on the walls of Hamidia Hospital and elsewhere for several weeks, were never reclaimed by their families. Number 436 was a young woman with tattoos on her cheeks; 213 was an emaciated old man with long white hair; 611 was an adolescent with a bandaged forehead; 612 a baby only a few months old. Who were these people? We will never know.

It is now estimated that the gas from the 'beautiful plant' killed between 16,000 and 30,000 people.

More than half a million Bhopalis suffered from the effects of the toxic cloud, in other words three in every four inhabitants of the city.[1] After the eyes and lungs, the organs

---

[1] Exactly 521,262 people according to the Indian Medical Research Council. This figure does not include victims who were not permanent residents of Bhopal, all those of 'no fixed abode' or members of nomadic communities. Nor does it include those victims indirectly affected by the tragedy, such as children still in their mothers' wombs, or those subsequently born to parents poisoned by the gas.

most affected were the brain, muscles, bone joints, liver, kidneys, reproductive as well as the nervous and immune systems. Many of the victims sank into such a state of exhaustion that movement became impossible. Many suffered from cramps, unbearable itching or repeated migraines. In the bustees women could not light their *chulas* to cook food without risk of the smoke setting off pulmonary haemorrhaging. Two weeks after the accident, a jaundice epidemic struck thousands of survivors who had lost their immune system defences. In many instances neurological attacks caused convulsions, paralysis, and sometimes coma and death.

More difficult to assess, but just as severe, were the psychological consequences. In the months that followed the disaster, a new symptom made its appearance. The doctors called it 'compensatory neurosis'. A number of Bhopalis developed imaginary illnesses. But some neuroses were very real. The most serious psychological effect was *ghabrahat*, a panic syndrome which plunged patients into a state of uncontrollable anxiety. With an accelerated heartbeat, sweating and shaking, those suffering from it lived in a permanent nightmare state. People with a tendency towards vertigo suddenly saw themselves on the edge of a precipice; those who were frightened of water thought they were drowning. With its associated depression, impotence and anorexia, *ghabrahat* brought desolation to a large number of survivors, sometimes making them view the catastrophe as a divine punishment, or as a curse inflicted on them by some member of their family. *Ghabrahat* drove many to despair and suicide.

Today Bhopal has some 150,000 people chronically affected by the tragedy, which still kills ten to fifteen patients a month. Breathing difficulties, persistent coughs, ulcerations of the cornea, early-age cataracts, anorexia, recurrent fevers, burning of the skin, weakness and depression are still manifesting themselves, not to mention constant outbreaks of cancer and tuberculosis. Chronic gynaecological disorders such as the absence of periods or, alternatively, an increase to four or five times a month, are common. Finally, retarded growth has been noted in young people aged between fourteen and eighteen, who look scarcely ten. Because Carbide never revealed the exact composition of the toxic cloud, to this day medical authorities have been unable to come up with an effective course of treatment. Thus far, all treatments have produced only temporary relief. Often overuse of steroids, antibiotics and anxiolytics serves only to exacerbate the damage done by the gases. Today Bhopal has as many hospital beds as a large British city. Without enough qualified doctors and technicians to use and repair the ultramodern equipment, however, the vast hospitals built since the disaster remain largely unused. An enquiry carried out in July 2000 revealed that a quarter of the medicines dispensed by the Bhopal Memorial Hospital Trust, recently established by Carbide, were either harmful or ineffective, and that 7.6 per cent were both harmful and ineffective.

So much official negligence produced a rush of new private medical practices. According to victims' advocacy groups, however, two thirds of these doctors lack the necessary skills. In light of this, several of these groups set up their own care centres such as the Sambhavna Clinic, with which the authors of this book are now involved. This unique institution, founded by a former engineer (see the *Letter to the Reader*) by the name of Satinath Sarangi, is staffed by four doctors, some twenty medical and welfare experts and several foreign volunteers. Together they monitor more than 10,000 economically disadvantaged patients, and see that they all receive effective treatment. The team at Sambhavna Clinic has experienced that certain yoga exercises can dramatically improve chronic respiratory problems. Half the patients thus treated have regained the ability to breathe almost normally and have been able to give up the drugs they had been taking for many years. The clinic also manufactures some sixty plant-based Ayurvedic medicines, which have already enabled hundreds of patients to resume some form of activity – a spectacular achievement that has wrested from poverty some of the 50,000 men and women once too weak to do manual work.

So many years after the catastrophe, 5000 families in Chola, Shakti Nagar, Jai Prakash Nagar and other bustees are still drinking water from wells polluted by the waste that once came from the factory. Samples taken by a Greenpeace team in December 1999 from the vicinity of the former installation showed a carbon tetrachloride level 682 times higher than the acceptable maximum, chloroform level 260 times higher, and a trichloroethylene level 50 times higher.

❦

No court of law ever passed judgement on Union Carbide for the crime it committed in Bhopal. Neither the Indian government, claiming to represent the victims, nor the American lawyers who had extracted thousands of powers of attorney from poor people like Ganga Ram, managed to induce the law on the other side of the Atlantic to declare itself competent to try a catastrophe that had occurred outside the United States. Yet one of the American lawyers representing the Indian government had taken young Sunil Kumar, one of the three survivors of the family of ten children, to New York, to try and persuade the judge in the New York Court before whom the case had been brought, to agree to try Carbide. It was the 'ambulance chaser's' view that only an American court could require the multinational to pay an amount commensurate to the enormity of the wrong. They sought damages of up to 15 billion dollars. Carbide's defence lawyers argued that an American court was not competent to assess the value of a human life in the Third World. 'How can one determine the damage inflicted on people who live in shacks?' asked one member of the legal team. One newspaper took it upon itself to do the arithmetic. 'An American life is worth approximately five hundred thousand dollars,' wrote the *Wall Street Journal*. 'Taking into account the fact that India's gross national product is 1.7 per cent of that of the United States, the Court should compensate for the decease of each Indian victim proportionately, that is to say with eight thousand, five hundred dollars[2].' One year after the catastrophe, no substantial help from the multinational had reached the victims, despite the fact that Carbide had given 5 million dollars in emergency aid. It took four long years of haggling before, in the absence of a proper trial, a settlement was drawn up between the American company and the Indian government. In February 1989, Union Carbide offered to pay 470 million dollars in compensation, in full and final settlement, and provided the Indian government undertook not to pursue any further legal proceedings against the company or its Chairman. This was over six times less than the compensation initially claimed by the Indian government. The lawyers for the government nevertheless accepted the proposal without consultation with the victims.

This settlement that Carbide could never have hoped for

---

[2] *Averting a Bhopal legal disaster, Wall Street Journal, 16 May 1985*

sent its stock up two dollars on Wall Street, a rise which enabled Chairman Warren Anderson to inform his shareholders that in the final analysis, the Bhopal disaster meant 'a loss of forty-eight cents a share' to the company. One week after the fateful night, Union Carbide shares had dropped fifteen points, reducing the multinational's value by 600 million dollars.

Most surprising was the psychological shockwave that the disaster triggered throughout every level of the company, from engineers like Warren Woomer or Ranjit Dutta, to ordinary workers, office employees or even liftboys in the various subsidiaries. At the head office in Danbury, secretaries burst into tears over telexes from Bhopal. Engineers, unable to comprehend what could possibly have happened, shut themselves away in their offices to pray. Local psychiatrists had employees of one of the world's largest industrial companies come pouring in, in a pitiful state of depression and bewilderment. Many admitted to having lost confidence in 'Carbide's strong corporate identity'. There were similar reactions in Great Britain, Ghana, and Puerto Rico, wherever, in fact, the flag with the blue logo was flying. Four days after the catastrophe, at midday on 6 December, over 110,000 employees at the 700 factories and laboratories stopped work for ten minutes 'to express our grief and solidarity with the victims of the accident in Bhopal'.

Anderson was so concerned by the crisis in morale of 'Carbiders' the world over that he recorded a series of video messages intended to restore confidence. These messages talked a great deal about ethics, morality, duty and compassion. The best way of getting things back on track, however, was still to show that the company was not guilty. On 15 March 1985, the Vice-President of the agricultural division of the Indian subsidiary, K. S. Kamdar, called a press conference in Bombay to announce that the tragedy had not

371

been due to an accident but to sabotage. Kamdar based his statement on the enquiry carried out by the team of engineers sent to Bhopal the day after the disaster. According to this enquiry, a worker had deliberately introduced a large quantity of water into the piping connected to the tank full of MIC. This worker, who remained nameless, had supposedly acted out of vengeance after a disagreement with his superiors. To support this theory, the investigators had relied on the discovery of a hose close to a tank, and in particular, upon the doctoring of logbook entries made by the shift on duty that night. The report that supposedly incriminated a saboteur made no mention of the fact that none of the factory's security systems were functioning at the time of the accident.

The authors of this book found the man Union Carbide had accused and talked to him at length. The man in question is Mohan L. Varma, the young operator who, on the night of the disaster, identified the smell of MIC while his companions attributed it to the fly spray Flytox used in the canteen. It is their deep-seated conviction that this father of three children, who was well aware of the dangers of methyl isocyanate, could not have perpetrated an act to which he himself and a large number of his colleagues were likely to fall victim. His colleague, T. R. Chouhan, wrote a book *Bhopal – The Inside Story*, pointing out large technical holes in Carbide's sabotage story. Mohan L. Varma's innocence was, moreover, immediately recognized. No legal proceedings were ever instituted against him. Today he lives quite openly two hours by car from Bhopal. If the survivors of the tragedy had had the slightest suspicion about him, would they not have sought vengeance? As it was, no one in Bhopal or elsewhere took the charge seriously.

Events would further conspire to refute it. Four months after the accident in Bhopal, on 28 March 1985, a methyl oxide leak at the Institute site in the United States poisoned eight workers. On the following 11 August, again at Institute, another leak, this time from a tank holding aldicarb oxime, injured 135 victims in the Kanawha Valley. One of them was Pamela Nixon, the laboratory assistant at Saint Francis Hospital in Charleston, who had noticed the smell of boiled cabbage years before. 'I was among those who believed Union Carbide when they claimed that accidents like the one in Bhopal could not occur in America,' she told the press when she came out of hospital. The incident had changed her life. She went back to college and joined the organization 'People Concerned About MIC', created by residents in her area. After which, armed with a degree in environmental sciences, she set out to take on the directors of the various chemical factories in the Kanawha Valley and compel them to tighten their safety measures. This was something that no one had done in Bhopal. The tragedy was bearing its first positive fruits.

❈

In Bhopal too, the victims organized themselves in pursuit of what was rightfully theirs. Activists' organizations rallying thousands of survivors, ransacked Carbide's offices in New Delhi and demanded the immediate payment of the promised indemnities. Five years after the tragedy, its victims had still not laid hands on a single one of the 470 million dollars they had been awarded.

Not surprisingly, so large a sum of money, even though placed in a special account administered by the Supreme Court, was a magnet for the greedy. Sheela Nadar, Padmini's mother, had to pay out 1400 rupees for a dossier establishing her husband's death. Payment of backsheesh became obligatory in order to obtain access to the compensation

373

pay desks or to the often very distant offices that handed out the first allocations of provisions and medical aid. In the final analysis, according to official figures, about half a million survivors would eventually receive what was left of the money paid by Carbide: a little less than 60,000 rupees or approximately 1000 pounds for the death of a parent, and about half that in cases of serious personal injury. It was a far cry from the million rupees the New York lawyer had promised Ganga Ram and the Orya Bustee survivors.

❊

Because the wind had been blowing in the direction of the bustees that night, it was the poorest of the poor who were most affected by the tragedy. Left to suffer, exploited by predators on all sides, the survivors soon found themselves subject to further persecution. Under the guise of a 'Beautification Programme' the new authorities, who had risen to power in 1990, diverted moneys meant for the victims into emptying the bustees of any Muslims. Flanked by police, bulldozers razed several neighbourhoods to the ground. Only the determination of about fifty Muslim women threatening to burn themselves to death succeeded in putting a halt to the eviction of Muslims. But after a few days, they were all moved to Gandhinagar outside the city. Iqbal, Ahmed Bassi and Salar who escaped the scourge of MIC, were driven out by the madness of men. Like most of the other Muslims living in the Kali Grounds' neighbourhood, they had to abandon their homes again. This time for good.

❊

In 1991, the Bhopal court summoned Warren Anderson, Union Carbide's Chairman, to appear on a charge of 'homicide in a criminal case'. But the man who was enjoying peaceful retirement in his villa in Vero Beach, Florida, did not keep the promise he had made to a journalist as he left Indian soil on 11 December 1984. Not only was he not returning to the country where his company had wrought disaster, but he actually managed to disappear within his own country. Anderson left Vero Beach, and no one knows to this day where he found refuge. The international warrant for his arrest issued under Indian law via Interpol remained unserved. In March 2000, in response to a class action suit by victims' organizations in the Federal Court of the southern district of New York, Union Carbide's lawyer William Krohley said the company will accept process served in the name of Anderson but will not disclose his whereabouts. These organizations remain undaunted, however, and do not intend to give up. The graffiti 'HANG ANDERSON', which the survivors never tire of repainting on their city walls, are a reminder that justice has not yet been done.

If Warren Anderson is a fugitive from the Indian courts, the prospects of bringing Union Carbide to justice are just as unlikely, for the very good reason – albeit one of small consolation to the victims – that the multinational no longer exists as such. Despite all its Chairman's efforts, the tragedy on 2 December 1984 signed the death of the proud company with the blue and white logo. The purchase of its agricultural division first by the French company Rhône-Poulenc, then by Bayer, now the proprietor of the Institute Sevin factory, and the takeover in August 1999 of all of its assets, for the sum of nine billion, three hundred million dollars, by the Dow Chemical group, has forever evicted Union Carbide from the world's industrial horizon. The initiators of the various legal proceedings launched against the Danbury multinational let it be known that they would hold Dow Chemical legally responsible for the charges against Carbide. Their claim was given short shrift by Dow's President: 'It is not in my power,' declared Frank Popoff, 'to take responsibility for an event which happened fifteen years ago, with a product we never developed, at a location where we never operated.'

❧

And what of the 'beautiful plant'? One day in January 1985, shortly after *Operation Faith*, a *tharagar* turned up outside the tea house in Orya Bustee.

'I'm looking for hands to dismantle the rails from the railway line leading to the factory,' he said.

The stretch of track linking the factory to the main railway line had never been used. It was a testimony to the megalomania of the South Charleston engineers who purchased both a locomotive and freight cars, to transport the enormous quantities of Sevin the factory was supposed to produce. Timidly, Ganga Ram, who had lost most of the customers of his painting business in the catastrophe, put up his hand.

'I'm looking for work,' he said, convinced that the *tharagar* would reject him when he saw his mutilated stumps.

But that day Carbide was taking on any available hands. The former leper would at last be able to have his revenge by helping to dismantle the monster that had once refused him employment.

For one year, Jagannathan Mukund, the last managing director, headed the team assigned to closing down the factory, a Herculean task which involved cleaning every piece of equipment, every pipe, every drum and tank, first with water and then with a chemical decontaminant. These cleaned and scrubbed parts of the factory were sold off to smal local entrepreneurs. In 1986, when the job was done, the last workmen wearing the prestigious boiler suit with the blue and white logo left the site for ever.

Today, the abandoned factory looks like the vestige of some lost civilization. Its metal structures rust in the open air. In the rough grass lie pieces of the sarcophagi that protected the tanks. On the control room walls, the seventy dials rest in eternal peace, including the pressure gauge for tank 610 with its needle stuck on the extreme left of the instrument, lasting testament to the fury of MIC. The notices with the inscription 'SAFETY FIRST' add a touch of irony to a scene of industrial devastation.

What was to be done with this mute but powerful witness? In 1997, local municipal authorities suggested turning the whole of the Kali Grounds site into an amusement park. But the indignant outcry the proposal provoked caused the authorities to withdraw it. The accursed factory would remain there always, as a place of remembrance.

※

Fortunately, a privileged few managed to escape the misfortune that befell most of the tragedy's victims. Orya Bustee's bride and groom were among them. Miraculously resurrected after her rescue from the funeral pyre, Padmini was able to rejoin her loved ones after a long and painful stay in Hamidia Hospital. She returned to Orya Bustee and set up home with her husband Dilip in her parents' hut. Very soon, however, the nightmare of that tragic night began to haunt her to the point where she could no longer bear the place in which she had spent her adolescence. The mere sight of the metal structures mocking her from a few hundred yards away very nearly drove her insane. That was when an opportunity presented itself in the form of a plot of land for sale, about forty miles from Bhopal, almost on the banks of the Narmada River. The idea of returning along the trail that had once brought her family from Orissa to Bhopal filled the young Adivasi with enthusiasm. She persuaded her husband that they could make their home in this rural area, have a small farm and live on what they produced. Her mother and brother were prepared to go with them. The indemnity they had just received for the death of the head of their family made the relocation just about feasible.

Dilip and Padmini built a hut, planted soya, lentils, vegetables and fruit trees. Little by little they dug out an irrigation system. Like all the peasants in the area, they bought 'medicines' from travelling salesmen to protect their crops from insects, especially from the weevils that liked to attack potatoes. These door-to-door salesmen did not, of course, carry Sevin. Instead they had pyrethrum-based pesticides, which had the advantage of being both cheap and generally effective, except when it came to soya caterpillars, which were a real nightmare.

One day in the autumn of 1998, Dilip and Padmini received a visit from a pesticide salesman they had never seen before. He was wearing a blue linen boiler suit with a badge on it. Padmini, who, thanks to Sister Felicity, had learned to read, had no difficulty in making out the name on the badge. It was that of one of the giants of the world's chemical industry.

'I'm a Monsanto rep,' he declared, 'and I've come to give you a present.'

With these words, the man took out of his motorized three-wheeler a small bagful of black seeds which he proceeded to place in Dilip's hands. 'These soya seeds have been specially modified,' he explained. 'They contain proteins, which enable them to defend themselves against all kinds of insects, including caterpillars . . .' – Seeing that his audience was wide-eyed with interest, the man seized his opportunity: 'I can also offer you sweet pepper seed that's immunized against plant-lice, lucern seed treated against diseases affecting cows, sweet potatoes that . . .'

Their benefactor had brought the Indian peasant couple a whole catalogue of miraculous products. All the same, there was nothing charitable about his visit. It was the result of a marketing campaign thought up some 13,000 miles away, in California where Monsanto, leader in the latest biotechnical revolution, had its head office. Thirty years after Eduardo Muñoz and his Sevin, it was Monsanto's turn to take an interest in the Indian market.

Padmini took the bag of seeds and went and placed them on the small altar with its image of Jagannath she had set up in the entrance to the hut, just next to a *tulsi* tree. Dilip and she would wait for the end of the monsoon to plant the little black granules. Of course, neither of them was aware that these marvellous little seeds had been genetically engineered not to reproduce. The soya they harvested would not supply the seeds for another crop. As to the health risks this transgenic engineering might represent, neither the Monsanto sales representative nor his new customers would even begin to think about them. Wasn't India the perfect place for a new generation of sorcerer's apprentices to conduct their experiments? If everything the salesman had told them was true, Padmini and Dilip were quite sure that their lives were going to change for ever. They could burn incense to thank their god. For the future belonged to them.

# What became of them

**Warren Anderson** – Chairman of Union Carbide at the time of the tragedy, Anderson left the company in 1986 and retired to Vero Beach, Florida. Following complaints filed against him by the victims' organizations and an international Interpol warrant, he disappeared from his home address.

**Shyam Babu** – The restaurateur who had promised to 'feed the whole city' and who supplied the wood for the cremations still presides over the till in his restaurant. His business has expanded with the opening of a four-storey hotel above the Agarwal Poori Bhandar. At thirty rupees a room, Shyam Babu's rates are still unbeatable.

**Sajda Bano** – The widow of Mohammed Ashraf, the factory's first victim, is yet to receive compensation for her husband's death. She is fighting to collect what is still due to her for the death of her eldest son, Arshad. Soeb, the younger son, is suffering from serious neurological and other disorders as a consequence of the catastrophe. Both live on the ground floor of a small cottage next to the 'widows' colony'. Both Sajda Bano and Soeb are treated in the Sambhavna Clinic that houses the gynaecology clinic set up by Dominique Lapierre.

**John Luke Couvaras** – The engineer whose wife was massaged by eunuchs, has nostalgic memories of those splendid days when he helped to build the 'beautiful plant'. He is now living in Greece but dreams of building a house on the sacred banks of the Narmada.

**Suman Dey** – The chief engineer on duty in the factory control room on the night of second to third of December, set up a mechanical fabrication unit with the severance pay he received from Carbide. His mini-factory is on the verge of closing down owing to business losses.

**Sharda Diwedi** – The Managing Director of the power station that supplied the lighting for the weddings on the fateful night retired in Bhopal. He suffers from chronic shortness of breath, which he attributes to his efforts to save the guests at the wedding of his niece Rinou, whose marriage could only be celebrated several days after the catastrophe. Ten years later, Rinou's husband died of a cancer that the Diwedis see as a consequence of poisoning by the toxic cloud. As for Rinou, she suffers from recurrent bouts of depression. The catastrophe destroyed her life.

**Ranjit Dutta** – The Indian engineer who, along with Eduardo Muñoz, built the first Sevin formulation factory and who tried, four months before the accident, to alert his superiors to the dilapidated state of the factory, retired in Bhopal. He works as a pesticide consultant for several chemical manufacturers.

**Sister Felicity** – The Scottish nun who saved dozens of children on the night of the catastrophe and over the ensuing days, still runs the House of Hope, which takes in children with physical and mental handicaps.

**Dr Deepak Gandhe** – The doctor on duty at Hamidia Hospital on the night of the disaster, left Bhopal to open a practice in the small town of Kandhwa, on the route to Bombay. He devotes part of his time to humanitarian work in the poor areas of Bihar.

**Rajkumar Keswani** – The Cassandra who predicted the catastrophe in his newspaper now works as a reporter for New Delhi television. He did not profit from the far-sighted articles that made him, for a while, India's most famous journalist.

**Rehman Khan** – The poetry-loving factory worker, who became an instrument of destiny, still lives in Bhopal. He works for Madhya Pradesh's forestry department.

**Colonel Khanuja** – The Sikh officer whose family were murdered while returning from a pilgrimage to Amritsar, and who, on the night of the disaster, wrested hundreds of inhabitants of the poor neighbourhoods near the Carbide factory from the gas, is now living in Jaipur. Ever since the fateful night, he has had breathing difficulties and is gradually losing his sight. In 1996, he tried to obtain financial assistance from Carbide to go to the United States for an eye operation that Indian specialists are unable to perform. Despondent at the prospect of becoming completely blind, this hero of that tragic night is still waiting for a response.

**Professor N. P. Mishra** – The dean of the medical college who roused all the faculty students from their beds, telephoned all Madhya Pradesh's pharmacists and arranged for emergency aid, is still Bhopal's leading medical authority. He sees patients in his superb villa on Shamla Hill, plastered with diplomas and distinctions awarded by medical institutions all over the world. A notice displays the price of consultation: one hundred and fifty rupees, approximately two pounds.

**Jagannathan Mukund** – Following the closure of the Kali Parade plant, the factory's last managing director left Bhopal to live in Bombay where, for several years, he went on working for Union Carbide. He retired to Karnataka, a southern state. He is still under indictment by an Indian court to stand trial for his role in the tragedy.

**Eduardo Muñoz** – After running Union Carbide's agricultural products division for several years, the flamboyant Argentinian engineer who fathered the Bhopal factory, moved to San Francisco where he now sells wine chiller cabinets.

**Padmini** and her husband **Dilip** – See the *Epilogue*

**Kamal Pareek** – The Indian engineer who left his 'beautiful plant' because he could not bear to see its safety standards declining, now lives in New Delhi where he works as an independent consultant to the chemical industry.

**Shekil Qureshi** – The Muslim supervisor, who was the last to leave the factory on the night of the catastrophe, now runs a factory for production of alum used in purifying water. He is suffering from serious respiratory after-effects. Like Mukund, he too is under indictment to stand trial for his role in the tragedy.

**Ganga Ram** – The leprosy and gas survivor has his small house painting business running again. The Bhopal municipal government gave the occupants of Orya Bustee a plot of land less than a mile north of the Kali Grounds. The community settled there and has reconstructed a small, typically Orya village with mud huts decorated with geometric designs. **Dalima** is still very active, although she complains more and more of the effects of the severe fractures to her legs.

**Dr Sarkar** – The heroic railway doctor was found at death's door in the stationmaster's office. Since then he has suffered from a chronic cough and frequent attacks of suffocation. For years, he was convinced that pockets of gas left behind by the toxic cloud were still poisoning people. He retired in Bhopal, where he lives surrounded by his children.

**Dr Satpathy** – The rose enthusiast forensic expert who performed the first autopsies on the victims on the night of the tragedy, is now head of the Department of Pathology at the Gandhi Medical College in Bhopal. He still grows roses, which he sends to all the Indian flower shows. Affected by the gases that had impregnated the clothing on the corpses, he now suffers from breathing difficulties. Because he did not live in the area hit by the toxic cloud, he never received any compensation.

**V. K. Sherma** – The courageous deputy stationmaster who saved hundreds of passengers by making the Gorakhpur Express leave the Bhopal station, now lives in the suburbs of Bhopal. His injuries have turned him into an almost total invalid. His breathing is so laboured that he can scarcely speak. The slightest physical effort causes terrible attacks of suffocation. The government paid him 35,000 rupees, a little over 400 pounds' compensation.

**Arjun Singh** – The Chief Minister of Madhya Pradesh who dispensed property deeds to the occupants of the poor neighbourhoods bordering on the Union Carbide factory, won the elections in February 1985 and became one of India's most powerful politicians. Appointed Vice-President of the Congress Party by Rajiv Gandhi, he was made a government minister several times. He has lost his seat in the New Delhi parliament. He divides his time between the capital and Bhopal where he has had a sumptuous residence built on the shores of the Upper Lake.

**Mohan Lal Varma** – The operator, accused of sabotage by Union Carbide, was never charged. Today he works for Madhya Pradesh's Industries Department.

**Warren Woomer** – The American engineer who supervised the training of the 'beautiful plant's' Indian engineers at Institute, is now living with his wife Betty in South Charleston. His house overlooks the Kanawha Valley. When Woomer goes out for a walk, he can see the outline of the Institute factory, where the tanks invariably contain several dozen tons of methyl isocyanate. Woomer has just written a history of Union Carbide's industrial presence at Institute. He has remained a consultant for the factory, which now belongs to Bayer.

# 'All *that* is *not given* is *lost*'

## SOLIDARITY WORK THAT WE HAVE UNDERTAKEN IN CALCUTTA, RURAL BENGAL, GANGES DELTA, MADRAS AND BHOPAL

Thanks to royalties and my fees as a writer, journalist and lecturer, and thanks to the generosity of my readers and friends who support the organization I founded in 1982, it has been possible to initiate or maintain the following humanitarian work:

1. The assumption of complete and continuing financial responsibility for taking care, at the Udayan-Resurrection home in Barrackpore near Calcutta, of three hundred young boys and girls who have suffered from leprosy.

2. The assumption of total and continuing financial responsibility for 125 handicapped children in the Mohitnagar and Maria Basti homes, near Jalpaiguri.

3. The construction and equipment of the Backwabari home for severely mentally and physically disabled children.

4. The expansion and reorganization of the Ekprantanagar home in a destitute suburb of Calcutta, which provides shelter for 140 children of seasonal workers at the brick kilns. The installation of a source of clean drinking water has transformed the living conditions in this home.

5. The creation of a school near the Ekprantanagar home to educate both the 140 children who live there and 350 very poor children from the nearby slums.

6. The reconstruction of several hundred huts for families who have lost everything in the cyclones that have hit the Ganges Delta.

7. The assumption of total financial responsibility for the Banghar SHIS medical centre and its programme to eradicate tuberculosis, which reaches out to more than two thousand villages. (Programme staff hold nearly 100,000 consultations annually.) The installation of X-ray equipment in the main dispensary and the creation of several subsidiary medical centres and mobile units providing diagnostic X-rays, vaccinations, medical treatment and nutritional care.

8. The establishment of four medical units in the isolated villages of the Ganges Delta, which provide vaccinations, treatment for tuberculosis, programmes in preventative medicine, patient education, family planning, as well as 'eye camps' to restore sight to patients with cataracts.

9. The sinking of tube wells for drinking water and the construction of latrines in several hundred villages in the Ganges Delta.

10. The launching of three floating dispensary-boats in the

Ganges Delta to bring medical aid to the isolated inhabitants of fifty-four islands.

11. In Belari, the assumption of financial responsibility for a rural medical centre which serves more than 90,000 patients a year from hamlets devoid of any medical care; the construction and assumption of responsibility for the ABC centre for physically and mentally handicapped children; the construction of a village for 100 destitute or abandoned mothers and children; with a home where mentally sick women are taken care of.

12. The creation of, and assumption of total responsibility for, several schools and allopathic and homoeopathic medical centres in two particularly poverty-stricken slums on the outskirts of Calcutta.

13. The construction of a 'City of Joy' village to house homeless tribal families.

14. The installation of solar-powered water pumps in ten very poor villages in the states of Bihar, Haryana, Rajasthan and Orissa, to enable the inhabitants to grow their crops even in the dry season.

15. The assumption of financial responsibility for a job-training workshop for leprosy sufferers in Orissa.

16. The provision of medicines as well as 70,000 high-protein meals for the children who live at the Udayan-Resurrection home.

17. Various undertakings for the underprivileged and leprosy patients in the state of Mysore; abandoned children in

Bombay, in Palsunda, near Bangladesh and in Rio de Janeiro, Brazil; as well as the occupants of a village in Guinea, Africa, and abandoned and seriously ill children in a hospital in Lublin, Poland.

18. The creation and financing of a gynaecological clinic in Bhopal to treat underprivileged women who are survivors and victims of the 1984 chemical disaster. The purchase of a colposcope to detect and treat cervical cancers.

19. The dispatching of emergency teams and aid to victims of the terrible floods in Orissa and Bengal; an on-going pro-gramme to house thousands of families who lost everything.

20. Since 1998, the assumption of financial responsibility for part of Pierre Ceyrac's education programme for 25,000 children in the Madras region.

# HOW YOU CAN HELP US
# TO CONTINUE OUR WORK AMONG SOME
# OF THE WORLD'S MOST UNDERPRIVILEGED MEN,
# WOMEN AND CHILDREN

Because of lack of resources, the association 'Action Aid for the Children of Lepers in Calcutta', which I founded in 1982, can no longer meet all the urgent needs, which the various Indian organizations that we have been supporting for the last twenty years, have to provide for.

In order to continue financing the homes, schools, clinics and development programmes run by the admirable men and women who have devoted their lives to serving the poorest of the poor, we need to find fresh support.

We have, furthermore, an ongoing serious worry. What would happen if tomorrow we were to have an accident or if illness were to prevent us from meeting the budgets for the centres that depend on us?

There is only one way to address this danger and that is to turn our association into a foundation.

The capital from this foundation would have to be able to provide the annual revenue necessary to finance the various humanitarian projects that we support. To generate the 500,000 dollars needed each year, we would need an initial capital sum of at least 10 million dollars.

How are we to raise that sort of capital if not through the contributions of a multitude of individuals?

Ten million is ten thousand times a thousand dollars. For some people it is relatively easy to give a thousand dollars to a good cause. Some people could probably give even more.

But for the vast majority of friends who have already spontaneously given us a donation after reading *City of Joy*,

*Beyond Love* and *A Thousand Suns* – or hearing one of my talks, and who often faithfully keep up their generous support, it is much too large a sum.

One thousand dollars, however, is also twice five hundred dollars or four times two hundred and fifty, or five times two hundred dollars, or ten times a hundred dollars, or even a hundred times ten dollars.

Such a sum can be raised from several people at one person's initiative. By photocopying this message, by spreading the word, by joining with other family members, friends or colleagues, by setting up a chain of compassion and sharing, anyone can help to keep this world alive and bring a little justice and love to the poorest of the poor. Alone we can do nothing but together all things are possible.

The smallest gifts count for just as much as the largest. Isn't the ocean made up of drops of water?

A big thank you in advance from the bottom of my heart for everyone's support, whatever their means.

D. Lapierre

Dominique Lapierre

P.S. We would like to remind readers that the association 'Action Aid for the Children of Lepers in Calcutta' has no administration costs. The totality of the money from the authors' royalties and of the donations received from readers is sent to the centres for which it is donated.

*'All that is not given is lost'*
Indian proverb

'ACTION POUR LES ENFANTS
DES LÉPREUX DE CALCUTTA'
(Action Aid for Lepers' Children of Calcutta)
Care of: Dominique & Dominique Lapierre
'Les Bignoles', Val de Rian, F-83350 Ramatuelle, France

Banking transfers can be made directly to:
Banque Nationale de Paris (BNP), Agency Paris Kléber
51, Avenue Kléber, F-75116 Paris, France
Bank Code: 30004    Agency Code: 00892
Account Number: 00001393127    Clé Rib: 21
IBAN: FR76 3000 4008 9200 0013 9312 721

For taxpayers in the UK, tax-deductible
contributions can be sent to:
'DOMINIQUE LAPIERRE CITY OF JOY AID,
ENGLAND'
c/o Kathryn Spink
Coachman's Cottage, Horsham Road
South Holmwood, Dorking, Surrey, RH5 4LZ

Banking transfers can be made
directly to the bank account:
Account Number: 61302914    Sort Code: 40-44-19
Bank: HSBC
Branch: 418 Ewell Road, Surbiton, Surrey KT6 7HJ
Tel: 020 8250 5250

*By saving a child,*
*by giving him the possibility to learn how to read and*
*write, by giving him a training,*
*it is the world of tomorrow that we save.*

- Rescuing, curing, feeding, clothing, schooling and training 10 leprosy-suffering children for one year costs £2000
- Digging one drinking water tube well in the saline areas of the Ganges delta costs an average of £300 to £1500
- Curing 100 TB patients costs about £1500

Dominique Lapierre's organization
has NO overhead costs.
Each donation received goes entirely to
serve a priority action.
See the website: www.cityofjoyaid.org

※

www.bhopal.org

The Sambhavna clinic's own website run by staff at the clinic in Bhopal and volunteers in the UK. Includes information on the gynaecology unit maintained by the Dominique Lapierre City of Joy Foundation; the combination of eastern and western therapies offered to survivors at Bhopal's only free clinic; survivors' testimonies and what you can do to help the victims of the world's largest industrial disaster. Also carries detailed information on the medical effects of exposure to MIC and research past and present.